Diving and Snorkeling Guide to

The Florida Keys

by John and Judy Halas
(Upper and Middle Keys)
Don Kincaid
(Lower Keys)
and the editors of Pisces Books

Pisces Books • New York

Publishers Note: At the time of publication of this book, all the information was determined to be as accurate as possible. However, when you use this guide, new construction may have changed land reference points, weather may have altered reef configurations, and some businesses may no longer be functioning. Your assistance in keeping future editions up-to-date will be greatly appreciated.
Also, please pay particular attention to the diver rating system in this book. Know your limits!

Second Printing 1987

Library of Congress Cataloging in Publication Data

Halas, John.
Diving and snorkeling guide to the Florida Keys.

Bibliography: p.
1. Skin diving—Florida—Florida Keys—Guide-books. 2. Scuba diving—Florida—Florida Keys—Guide-books. 3. Florida Keys (Fla.)—Description and travel—Guide-books. I. Halas, Judy. II. Kincaid, Don. III. Title.
GV840.S782U62 1984 797.2'3'0975941
84-1171

ISBN 0-86636-031-X

Staff

Publisher	**Herb Taylor**
Project Director	**Cora Taylor**
Series Editor	**Steve Blount**
Editors	**Carol Denby**
	Linda Weinraub
Assistant Editor	**Teresa Bonoan**
Art Director	**Richard Liu**
Artists	**Charlene Sison**
	Alton Cook
	Dan Kouw

Printed in Hong Kong

10 9 8 7 6 5 4 3 2

Table of Contents

How to Use this Guide

No matter what kind of diver you are, a trip to the Florida Keys will be one of the highlights of your diving memories. These little islands, some not much higher than two feet (60 centimeters) above sea level and the highest about eighteen feet (6 meters), average less than a half mile (1 kilometer) in width. Lying just offshore are the most beautiful underwater playgrounds in the United States. These tropical coral reefs extend from a point off the southern tip of the mainland and sweep south and west for 150 miles (240 kilometers) beyond Key West to the western limits of the Keys.

The Rating System for Divers and Dives

Our suggestions as to the minimum level of expertise required for any given dive should be taken in a conservative sense, keeping in mind the old adage about there being old divers and bold divers but few old bold divers. We consider a *novice* to be someone in decent physical condition, who has recently completed a basic certification diving course, or a certified diver who has not been diving recently or who has no experience in similar waters. We consider an *intermediate* to be a certified diver in excellent physical condition who has been diving actively for at least a year following a basic course, and who has been diving recently in similar waters. We consider an *advanced* diver to be someone who has completed an advanced certification diving course, has been diving recently in similar waters, and is in excellent physical condition. You will have to decide if you are capable of making any particular dive, depending on your level of training, recency of experience, and physical condition, as well as water conditions at the site. Remember that water conditions can change at any time, even during a dive.

The Christ of the Deep Statue in John Pennekamp Coral Reef State Park in Key Largo was the gift of an Italian diving equipment manufacturer, Egidi Cressi. It is a duplicate of a statue located in the Mediterranean off of Genoa, Italy. Photo: J. Halas. ▶

1

Overview of the Florida Keys

Long ago, the Florida Keys were coral patch reefs awash in the tropical waters of the warm southern sea. In time, the waters fell and the coral reefs were exposed to the sun and the winds. The corals died and the land grew. Offshore more corals formed and were, in turn, broken down by the sea. Finally, the exposed coral islands gained an identity and became known first as *Los Martires* and later as the Florida Keys.

Human beings came and peopled the shores of the islands, and though life was hard, vegetation sparse, and water scarce, the sea provided abundant provisions. Shellfish were plentiful and the early inhabitants left behind mounds of the remains of many meals. The Calusa Indians settled in villages in the Keys until they were victimized by the Spanish slavers and new diseases from another world, and this proud and powerful people were no more.

The Spanish were the next to leave their heritage with the little islands they called *cayos.* These colonists were dedicated to removing all the treasures of the New World and carrying them back to Spain. Their fleets of large merchant ships, laden with riches and protected by large armed galleons, sailed regularly from Havana northward through the Florida Straits just offshore of the Florida Keys. These ships, borne by the prevailing southeast winds, made Spain the richest and most powerful country in Europe. Yet the sea was never predictable and the violence of a fierce tropical storm could rise without warning, driving ships inshore, ripping out the bottoms of the vessels, and strewing their contents over the sands and coral. Today, dotting the reefs and bottom offshore, are still found the remains of ships of many countries driven into the shallows by hurricane winds or careless handling.

Most notable of the Spanish wrecks are the *Nuestra Señora de l'Atocha* and the *Margarita,* sunk in 1622 in the waters off Key West, and the galleons of the 1733 fleet blown onto the reefs of the Middle and Upper Keys. Millions of dollars in silver, gold, and artifacts have been recovered from these wrecks, and divers are still bringing up treasures buried for hundreds of years beneath the sea.

The power of the wind is still as popular a way to get around in the Keys as it was in the 16th, 17th and 18th centuries, when ships of the Spanish Plate Fleets, carrying gold and other treasures, regularly sailed these waters. Many of those ships wound up on area reefs. Their cannons, anchors and ballast rock can still be seen by divers. Photo: S. Blount. ▶

Naming the Keys

The names of many locations in the Keys are another Spanish legacy. Interestingly enough, some of the original Spanish names have provided the clues to lead researchers to the areas where shipwrecks were found. The Marquesas Islands, near Key West, were almost assuredly named for the Spanish Marquis, Marques de Cadereita, commander of the 1622 armada, who originally organized the search and salvage of the *Atocha* and her sister vessels. Making the connection between the place name and the historical event helped to determine the location and identity of the wreck site.

After the transfer of Spanish Florida to the United States in 1821, more and more people filtered into the Keys and commenced to make a home out of this ocean wilderness. They began an island trade and found that pineapples could be grown and bartered. Some lured ships onto the shallow reefs with misplaced lights and markers, then salvaged the wrecks. "Wrecking" became an established Keys business.

Many of these settlers were from the Bahamas, where many of their families, with Tory inclinations, had settled during the Revolutionary War. They returned to Key West, sometimes even bringing their homes with them piece by piece and reassembling them again. Some of these "conch houses" still stand today. These expatriates were called "Conchs" for the large pink mollusk which was their staple diet and is found in great quantities in the shallows offshore.

Islamorada, the "purple isle," stretches the length of Upper Matecumbe Key. This quaint town has many fine restaurants, established resorts, and a casual informal atmosphere all its own. The hurricane monument, toward the south end of town, commemorates the final resting place of many who lost their lives in the Labor Day Hurricane of 1935. Tea Table Bridge marks the end of Islamorada.

Lower Matecumbe Key is generally a residential area with some resorts and a marina facility at its southern end. Channel Two and Channel Five Bridges span the waters where the Keys begin a gradual bend to the west. Below this is the city of Layton and Long Key State Park. The Long Key Bridge was the second longest railroad viaduct in the Keys, and the old seal of the Florida East Coast Railroad showed a train heading over the concrete arches of this bridge.

An extension of North America into the tropical Caribbean, the Florida Keys contain the southernmost point in the United States, Key West. ▶

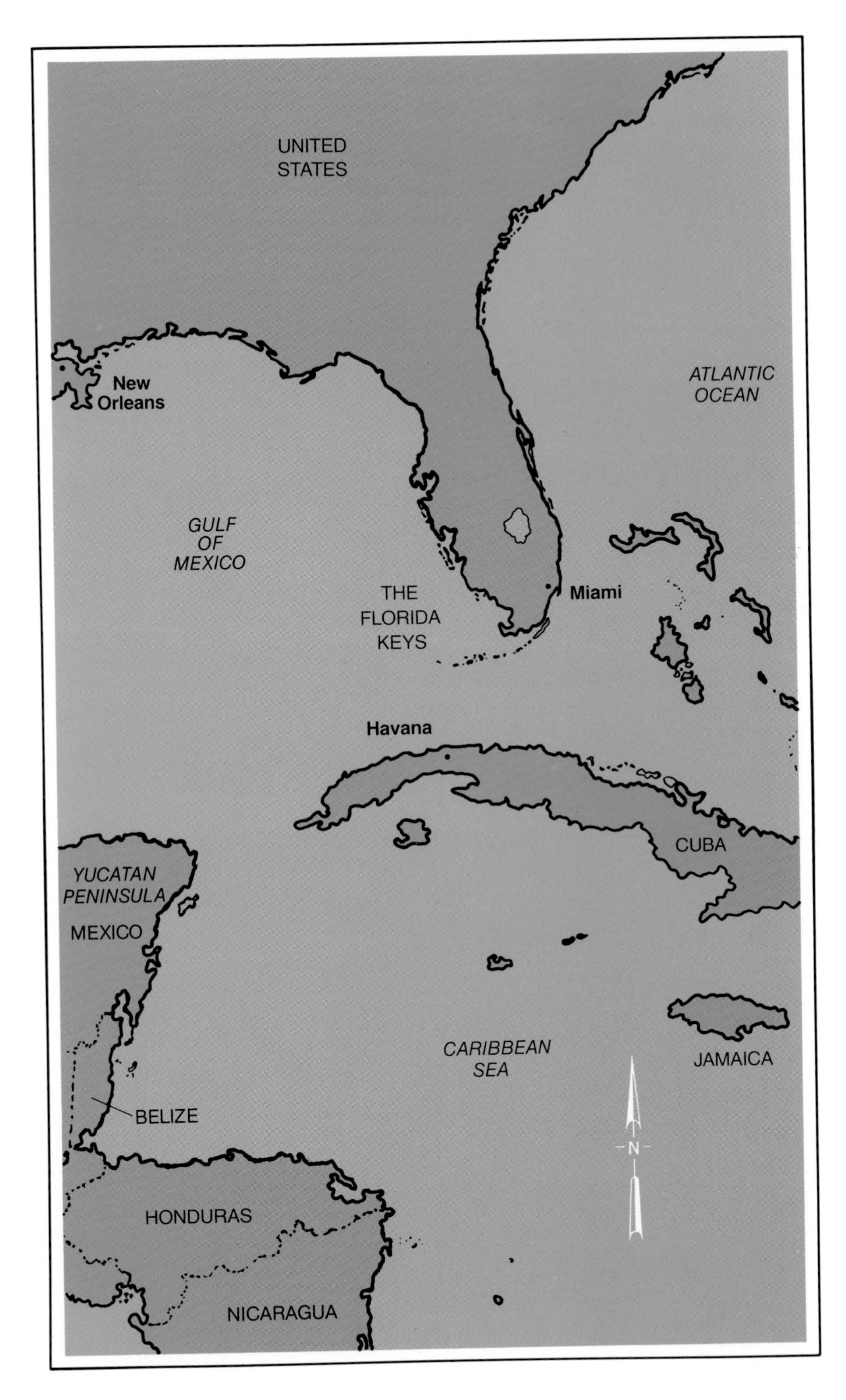
UNITED STATES
New Orleans
ATLANTIC OCEAN
GULF OF MEXICO
THE FLORIDA KEYS
Miami
Havana
CUBA
YUCATAN PENINSULA
MEXICO
CARIBBEAN SEA
JAMAICA
BELIZE
N
HONDURAS
NICARAGUA

The Keys Today

The Florida Keys have been touted as the "island diving you can drive to," and indeed it is true. The 135 mile (216 kilometer) journey down through the coral islands and across the sparkling turquoise waters is one of the most beautiful in the United States. Exiting the Florida Turnpike Extension at Florida City, visitors almost immediately enter an uninhabited stretch, populated only by birds and other wildlife, that separates the Keys from the Florida mainland. Travelers should look for the large osprey (fish eagle) nests built high on the telephone poles along the road. This expanse of mangrove and grassland swamp (essentially the same terrain that makes up the Everglades) ends at Jewfish Creek, the first of the Florida Keys's many bridges and the acknowledged entrance to the Keys.

Upper Keys. The road follows the old railroad embankment across Lake Surprise, so named because the railroad surveyors had no previous knowledge of the existence of this mile-wide saltwater lake. Rounding the bend onto Key Largo, the largest of the Keys's islands, the road becomes four-lane and continues southward the length of this Key.

The roadside is a crazy mix of businesses, resorts, residences, signs, and vegetation punctuated with an occasional high-rise condominium. The area of the Keys is mostly unincorporated and makes up the greater part of Monroe County. Locations are noted by milemarkers, small green signs along the road measuring the distance from Key West.

This cannon once belonged to a British warship, the Winchester, *which sank near Carysfort Reef in 1695. Photo: J. Halas.*

The queen angel fish is one of the most beautiful of the tropical species that grace the Keys. Photo: J. Halas.

The proliferation of dive shops on Key Largo is immediately apparent, and red and white "diver down" flags line either side of the highway. All these shops run dive trips into Pennekamp Park waters, but the actual entrance to John Pennekamp Coral Reef State Park/Key Largo National Marine Sanctuary is located near milemarker 103. The Key Largo area has an excellent selection of resorts, motels, campgrounds, restaurants, fast food outlets, and stores catering to the diving public. The community center is located at milemarker 100, where there are grocery stores, the post office, specialty shops, and other services.

The four-lane section of the highway ends at Tavernier on the southern end of Key Largo. A shopping center, housing among other businesses a supermarket and a movie theater, has recently been built here. Across the bridge at Tavernier Creek is the residential community of Plantation Key. The local high school, the community hospital, and the Sheriff's office are located in this area.

The Coast Guard station is at the south end of Plantation Key at Snake Creek bridge. This new bridge opens to permit tall vessels to pass through to either the Gulf waters or the Atlantic. More dive shops, restaurants, and resorts are located here and on Windley Key, just south of Snake Creek. Centered at the next bridge, Whale Harbor, is a large fleet of charter fishing boats where captains are happy to arrange fishing trips of just about any kind. Dive charters can be booked here as well.

2

Diving the Florida Keys

Stretching the length of the Florida Keys about six to seven miles (2 kilometers) offshore, the fragile Florida reef fringes the edge of the continental shelf. These reefs are the northern most living tropical coral reefs in the continental United States. Thousands of minute delicate animals congregate in living, growing communities that have grown into the amazingly varied structures of stony corals and the gently waving plumes, branches, and fans of the soft corals. The beauty of this underwater world has lured many to venture beneath the clear blue waters of the Atlantic to explore and behold the wonders it reveals.

Shore Diving. People who come to the Keys and want to sample the diving often ask where they can enter the water from the shore and see the coral reefs. Many are surprised to find that there are few actual reefs within swimming distance from shore. They also discover that there are no natural wide sandy beaches like those along the coast further north—the coral reefs offshore absorb the energy of the long waves that roll in from the sea and no force is left to break the rock into sand. The dynamic energy the corals need to thrive and grow diminishes closer to shore as well, and so they are not found either.

The island shores are fringed instead with mangroves, thick branching bushes that grow right out of the salt water. Their roots are like fingers that reach down and grasp the bottom to trap the debris which helps build the islands. In the water, hidden among these roots and branches, are small fish and other forms of sea life that take advantage of the offered protection until they are able to grow up and migrate to the sea.

Spreading out from the islands are the shallow flats. These are patches of turtle grass, sand, and some soft corals that provide the nursery grounds for the schools of juvenile fish that grow and populate the waters farther offshore. Without these safe havens, the schooling fish for which the Keys's reefs are known would not exist. Since few hard corals are found close to shore, the snorkeler who hopes to jump in the water and view the reef will find instead the fascination of a different world with rocks, mangrove channels, and grassy patches ready for exploration.

Snorkelers and divers can enjoy the many shallow coral areas all along the island chain. When water from the Gulfstream moves in over the Key reefs, visibility can be extraordinary. Photo: S. Blount. ▶

Most of the colorful reef inhabitants found throughout the Caribbean can be seen at one or another of the Keys dive areas. Photo: S. Blount.

The portion of the outer reef that is most commonly dived begins at a depth of thirty-five feet (11 meters) and extends back to the fifteen foot (5 meter) depth. This is also the region of greatest relief. Known as a "spur and groove system," sand gullies divide the coral-covered limestone ridges and generally run toward the sea. This dynamic section of the reef, assaulted by storm waves and bathed by the Gulfstream, provides an ideal habitat for marine organisms. A phenomenal variety of corals, fish, and other sea life thrive here and compete for space and food.

Above all, there is color. The hue of the water brightens the scene, and the spectacular brilliance of the tropical fish and the muted tones of the corals themselves blend together in technicolor richness. A close-up look reveals a sprinkling of other vivid highlights—tube worms, sponges, anemones, crustaceans, and more. It is little wonder that this is the favorite area for most of the diving in the Keys.

Boat Services

Rentals. People who have dived in the Keys before may want to rent a boat, and a number of places have rentals of various sizes available. Quality, equipment, and reliability vary, so ask to see the boats before making a decision. Rates are usually by the hour, half-day, and full day, but deposits are high and the renter has to take full responsibility for any damages.

Divers who decide to rent should be competent boat handlers and know how to read charts well. Someone who has no knowledge of the local waters should probably not plan to use this means of getting to the reef. Novice divers should also keep in mind that there will be no expert available to put them on the good reefs and watch out for their safety.

Dangers and Precautions. Captains are also financially liable for any damage they may cause to the corals, either by grounding or anchoring in the coral. The anchor should always be set in the sand and the chain should be free of the surrounding bottom. Whenever possible, boaters should make use of the mooring buoys that have been placed in Pennekamp Park and the Marine Sanctuaries.

All boats with divers in the water are required by Florida law to fly the "diver down" flag, red with a diagonal white stripe, and many charter boats now also fly the internationally recognized "Alpha" flag, a double pennant of blue and white. Both flags warn other boaters of divers in the area. Divers should remain within one hundred feet (30 meters) of their boats for these flags to serve their purpose. Boaters cruising through areas where there are divers should run at idle speed and look for divers' bubbles on the surface of the water.

A sailing catamaran draws up in a shallow, deserted lagoon in the Marquesas, southwest of Key West. Photo: D. Kincaid.

Dive Boats. By far, the greatest majority of visiting divers in the Keys choose to leave the driving to the professionals who run the charter dive boats. Any boat carrying passengers for hire must be run by a captain tested and licensed by the U.S. Coast Guard.

There are many dive shops in the Keys that offer diverse charters which take people out to the most popular reef sites. Most of the time, snorkelers and scuba divers are together on the same charter, so families and groups of people who do not all dive can go together. Occasionally this does mean that the snorkelers are in somewhat deeper water, rather than on the very shallow reefs where the diving would not be suitable for scuba. Some shops, however, do separate the two types of diving, and snorkelers are taken to a shallow, more protected, area.

The dive boats have varying capacities, but most shops have at least one boat certified to carry groups of twenty or more. Some shops may also have smaller boats that carry a maximum of six people. Special rates are sometimes available for groups and for people diving more than one day. Ask about these!

KADO. A majority of the dive shops and boat handlers in the Florida Keys belong to an organization known as KADO, the Keys Association of Dive Operators. This group was established to set standards of professionalism and diving safety in the Keys. KADO dive vessels carry crews trained in CPR, first aid, and the techniques of handling diving emergencies. Oxygen units and other safety equipment are available on board. Divers are required to be certified, have a logbook, and must also wear a buoyancy compensator and a pressure gauge.

When divers make arrangements to go on a charter, many dive shops in the Keys are now asking to see not only certification cards but dive logs as well. This is so they have a better idea about the experience of their passengers and can help them have an enjoyable and safe trip to the reef. Divers who have not been diving within the past year or who do not have a current logbook may be asked to go with a divemaster. Divers without much familiarity with ocean diving should make sure that the sea conditions are suitable for their level of expertise before going on a charter. They should let the boat's crew know if this is their first ocean dive so the crew can help them with unfamiliar procedures or give them additional advice.

Divers on a Keys charter trip usually are not accompanied by a crew member as a divemaster, as they often are at Caribbean resorts. Because the divers tend to spread out over the shallow reefs, it is easier to watch everyone at the same time from the boat. Any problems, which are uncommon, tend to be handled more easily from the surface. People who are uncomfortable in the water on their own should look for shops that provide a divemaster as an extra service for small groups who would like guided dives.

"Bug hunting" means just one thing in the Keys, diving for the succulent spiny lobster. Check the game regulations in the area you plan to dive before taking any animal—fish, coral or shells. Photo: J. Halas.

Current. The most important concern for divers in the Keys is the current. The Gulfstream flow has been clocked at four knots per hour, although along its edge over the reef it is usually much less. The outgoing tidal flow can also be strong. But the currents are deceptive. It is easy to be carried along with the flow and never notice the pull of the current until the time comes to swim back to the boat, low on air and energy. A diver in full scuba gear can have a very difficult time making progress with even a half-knot current. This is probably the most common diver safety problem for Keys divers and boat operators.

Underwater Hunting. You should be aware of the rules that govern the areas you dive in so you will know what is protected and if there are limits on your catch. Taking any form of coral is illegal anywhere along the Florida reefs. Lobster are protected during the closed season from April 1st through July 25th and may only be taken during the season by hand or net by divers. Undersized lobster with a carapace under 3 inches (1 centimeter) long and females bearing eggs may not be taken at any time. Severe penalties are in store for anyone molesting traps belonging to lobster fishermen.

Spearfishing is illegal within the state waters—inside the 3 mile (5 kilometer) limit—from Long Key north to Dade County. It is also not allowed near any public fishing piers or bridge catwalks. Spearfishing is especially prohibited within the protected areas of John Pennekamp Coral Reef State Park, the Key Largo National Marine Sanctuary, the Looe Key National Marine Sanctuary, and the Jefferson National Monument in the Dry Tortugas. Most of the marine life within the boundaries of these refuges is protected. It is better to leave things undisturbed and follow the diver's maxim of "take only memories and leave only bubbles!"

3

Diving the Upper Keys

John Pennekamp Coral Reef State Park and Key Largo National Marine Sanctuary. Sheltered by Key Largo, the largest of the Keys, the northernmost section of the reefline flourishes and some of the best developed reefs thrive here. Lush stands of magnificent elkhorn, huge mounds of star and brain corals, and shimmering soft corals are populated everywhere with myriads of fish and undersea creatures.

In the late 1950s, concerned citizens recognized that this beautiful and priceless resource was threatened by those sea-life collectors who would break, gather, and harvest the corals, shells, and fish for their own gain and not consider the consequences for the future. The concept of an underwater park established for the protection of the reefs gained momentum and by 1960, dedication ceremonies were held proclaiming the creation of the newest state park and the very first underwater marine park in the world. Known as John Pennekamp Coral Reef State Park, the name honors the Miami Herald newspaper editor who played a major role in sponsoring the creation of the park. By 1963 land acquisition and site construction were completed and Pennekamp Park was opened to the public.

This magnificent undertaking protected a 25 mile (40 kilometer) stretch of reefline from north of Carysfort Reef south to Molasses Reef and inshore to the coastline. The underwater area eastward from the edge of Florida, at 3 miles (5 kilometers), was granted by Presidential Proclamation in 1960 and rededicated as the Key Largo National Marine Sanctuary (KLNMS) under supervision of NOAA (National Oceanic and Atmospheric Administration) in 1975. This became the first marine sanctuary in a federal program that now oversees six such preserves around the country, including the Looe Key National Marine Sanctuary in the Lower Keys. Although it is still commonly known as Pennekamp Park, most of the offshore coral reef is actually protected and maintained by the KLNMS program. The combined park and sanctuary encompasses a total area of 183 square miles (75 square kilometers) and protects the waters out to the 300 foot (90 meter) depth.

Elkhorn coral often grows in high-energy zones where there is a rapid exchange of water. The older colonies often look like wizened oak trees, their spreading branches providing shelter for tropical fish. Photo: S. Blount. ▶

DIVE SITE RATINGS

UPPER KEYS	Novice Diver	Novice Diver w/ instructor or Divemaster	Intermediate Diver	Intermediate Diver w/ instructor or Divemaster	Advanced Diver	Advanced Diver w/ instructor or Divemaster
1 Christ of the Deep Statue*	×	×	×	×	×	×
2 Benwood Wreck	×	×	×	×	×	×
3 Carysfort Reef			×	×	×	×
4 Elbow	×	×	×	×	×	×
5 Grecian Rocks*	×	×	×	×	×	×
6 French Reef	×	×	×	×	×	×
7 Molasses Reef	×	×	×	×	×	×
8 White Bank Dry Rocks*	×	×	×	×	×	×
9 Pickles Reef	×	×	×	×	×	×
10 Conch Reef	×	×	×	×	×	×
11 Hen and Chickens	×	×	×	×	×	×
12 Keys Bridges				×	×	×
13 El Infante and the Spanish Treasure Ships	×	×	×	×	×	×

*Indicates Good Snorkeling Spot

When using the accompanying chart see the information on page 4 for an explanation of the diver rating system and site locations.

The dive shops of Key Largo, from Ocean Reef at the very north down to Tavernier and points south, regularly dive the reefs within the park/sanctuary. Three important navigational lights mark these reefs. Close to the northern edge of the park is Carysfort Light, named for a British ship that wrecked there in 1770. The Elbow is the middle light and marks a distinct bend, or elbow, in the reefline. At the south end is Molasses Light, one of the most popular and beautiful reefs in the Keys. Between these and inshore are a score of varied and fascinating dives, some well known and others virtually unexplored.

Lower Keys. The shops from lower Key Largo to Islamorada generally dive these reefs to the south. As a rule, the diving differs from that within the park, where dives are usually around depths of 35 feet (11 meters) or less and feature the spectacular coral development offshore of Key Largo. The shops further south often offer trips to deeper waters and may feature specialized dives like spearfishing or collecting.

If you wish to make a special dive, call or write the shops well in advance of your arrival to determine when a trip will be scheduled. Weather affects dive plans as well, so be prepared for rescheduling.

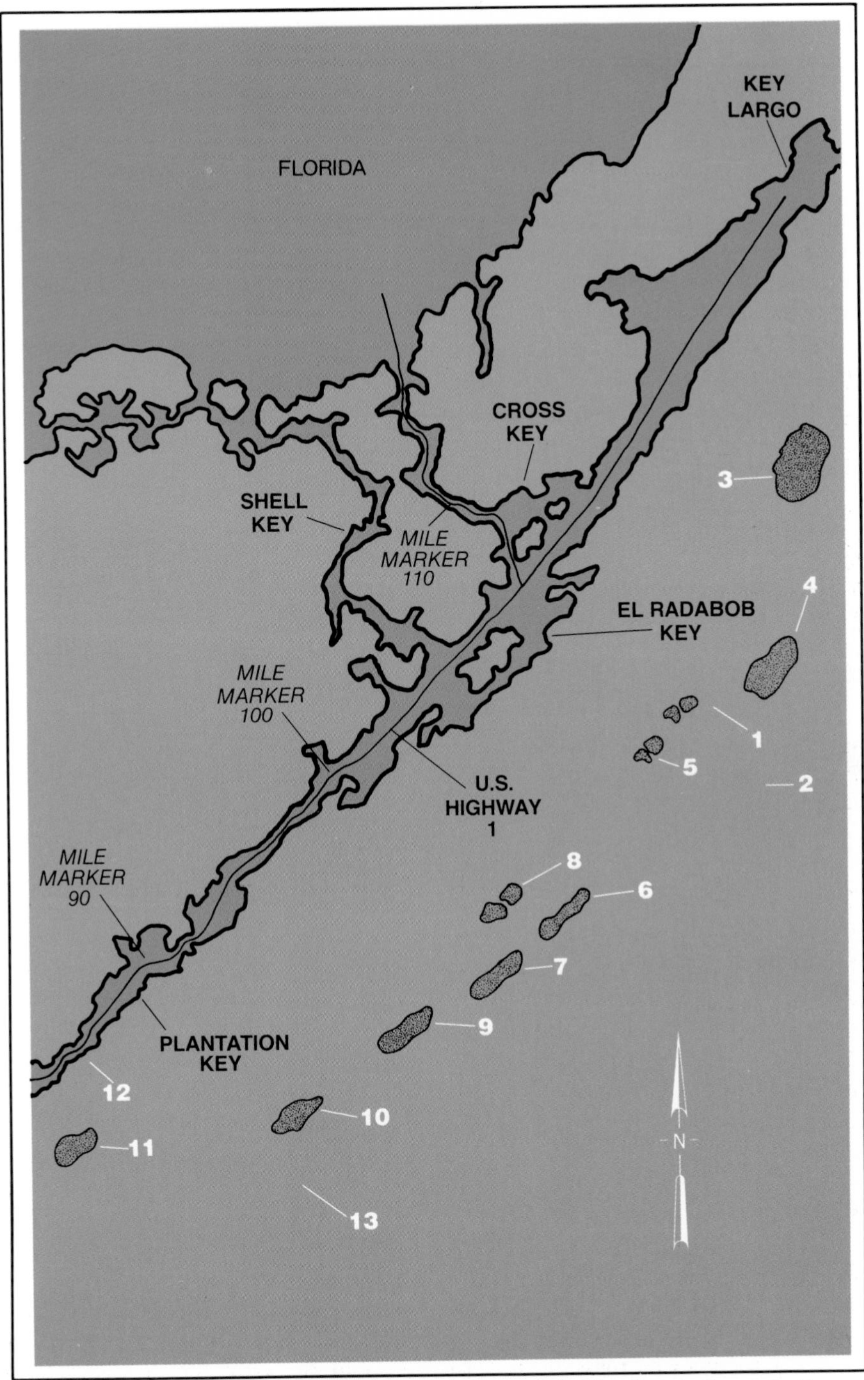

The Upper Keys stretch from Cardinal Sound in the north to Marathon to the south. Key Largo is the site of the first marine underwater park, John Pennekamp Coral Reef State Park, established in 1963.

The Christ of the Deep Statue* 1

Typical depth range	:	Shallow to 25 feet (8 meters)
Typical current conditions	:	Variable (none to moderate)
Expertise required	:	Novice
Access	:	Boat

The figure of Christ stands silhouetted against the blue waters of the ocean, His arms upraised to the surface, beckoning to the multitudes to enter and be with Him. This dramatic and memorable picture is one that most diving visitors to the Keys will want to carry home with them.

The Christ Statue was cast in Italy and donated to the Underwater Society of America by Egidi Cressi, an Italian industrialist and diving equipment manufacturer. It is a 9 foot (3 meter) tall bronze duplicate of the Christ of the Abysses statue, which stands in 50 feet (15 meters) of water off Genoa, Italy.

The Christ Statue is one of the most famous of Pennekamp Park's attractions. The 9-foot (3 meter) statue is cast of solid bronze. Photo: J. Halas.

This large brain coral head near the Christ Statue is noted for its size. Divers often find neon gobies stationed here, waiting to service other fish who wish to be cleaned of skin parasites by the gobies. Photo: J. Halas.

The Florida State Parks's bulletin about the statue has reprinted the original dedication of the Christ of the Abysses statue. It is beautiful and fitting:

> He will descend in the green silent depth of the sea and remain there to protect the living. The dead shall no longer be lonely. His carpet will be of soft algae; His naves and columns will be the pillars of the earth sunk in the great cerulean depths. Ships and phantoms of ships will crowd around Him; living men and dead men. The shadows of all those who lost their lives in the sea will be present, without discrimination of nationality, blood, or color. With His liberal gesture of invitation, He will welcome everybody; all who lived for the sea and who, in the same sea they so dearly loved, found their eternal peace.

The Christ Statue was set in place on the east side of Key Largo Dry Rocks, a beautiful reef area noted for its outstanding brain corals. Along the front side of the reef, the water is about 25 feet (8 meters) deep. The dive boats usually anchor or use the moorings on the offshore side where the bottom is generally sand, grass, and a few scattered rocks.

Near the statue, in the direction the left hand is pointing, is an extremely large brain coral noted for its size and symmetry.

Typical depth range : 50 feet (15 meters) offshore, bow section; 20 feet (6 meters) inshore, stern
Typical current conditions : Variable (none to strong)
Expertise required : Novice to intermediate
Access : Boat

The *Benwood* is one of the most outstanding wreck dives on the east coast and perfect as the first wreck dive for a novice diver. It is a large ship, easily accessible by boat, and it can be safely explored in a single dive. For those who are not used to dealing with the deceptive currents that tend to carry divers downstream, this wreck provides a focus; there is not much temptation to stray away from it since the bottom terrain is relatively barren in the surrounding area. The visibility is usually very good, but there is plenty of close-up viewing on days when it is less clear.

The wreck of the *Benwood* lies in line with the offshore reefs a mile and a half (two and one half kilometers) north of French Reef. Early in World War II, this freighter was torpedoed by a German submarine and then accidentally rammed by another vessel.

The ship lies with its bow offshore in 45 feet (14 meters) of water and its stern 300 feet (90 meters) inshore at a depth of 25 feet (8 meters). Begin your dive at the bow, which towers up from the bottom and makes a magnificent backdrop for a photographic panorama. The marine life growing across the broad expanse of this flattened surface provides good subject matter for close-up or macro photography.

The Benwood, *a modern steel wreck, has been broken into numerous pieces over the years. Its main attraction is the hordes of fish that swarm over the site. Photo: J. Halas.*

Through constant feeding, some of the Nassau groupers have become very friendly, even eating out of divers' hands. Photo: J. Halas.

Experienced divers might enjoy swimming out a short distance along the ledge that runs offshore from the bow. About 60 feet (18 meters) away, in 50 feet (15 meters) of water, the flukes of the ship's large anchor rise from the sand next to the ledge.

Before beginning to swim around the wreck, check for a current (watch the direction of your bubbles or rise and relax to see which way you float). Begin your dive along the downcurrent side of the wreck; most of the time you will be on the north side.

There are many places on the ship where you can swim under or through the wreckage, but no real inside exploration is possible. If you venture into any of the openings in the hull or under the twisted plates, be careful. Remember those spaces look larger than they are, and dangling hoses are easily caught on the jutting steel.

Inhabitants. Along the south side of the *Benwood*, look under the bottom of the hull for the moray eels, lobster, glassy sweepers and other fish that can be seen there. About midway along this side of the ship, you may want to venture out in the sand about 25 feet (8 meters) to look for the jawfish that nest there.

At the end of your dive, relocate your boat by looking up toward the surface. Because so many boats converge on this one popular place, boat traffic is very heavy here and you want to be sure to be near your boat when you surface. Try to ascend so you will come up near the dive platform and keep the current in mind if it is present. If you are running a boat in the area, keep a careful lookout for divers approaching the surface and be extremely careful when you place your anchor. Watch the bubbles!

Carysfort Reef 3

Typical depth range : 35–70 feet (11–21 meters)
Typical current conditions : Variable (none to moderate)
Expertise required : Intermediate
Access : Boat

Located at the extreme north end of Pennekamp Park/KLNMS, Carysfort Lighthouse marks the most remote regularly dived reef in the park. The site derives its name from the British vessel *HMS Carysfort,* which grounded here in 1770.

Most of the dive shops on the north end of Key Largo dive this reef, but it is usually a special trip rather than a regularly scheduled daily run. If you want to dive this area, it is best to check with various shops to find out when they plan to go. The one good factor about the relative inaccessibility of Carysfort Reef is that there is little boat traffic, even on busy holidays, so most of the time divers will have the reef to themselves.

Carysfort Reef has an unusual "double reef" configuration. There is a proliferation of shallow corals quite near the light, but these thin out as the depth gradually increases. A limestone bottom, dusted with sand and covered with sparse sponges and soft corals, extends out to about 35 feet (11 meters) where there is quite a steep drop down to 65 feet (20 meters). Along this "near wall" is a lush growth of coral colonies; staghorn corals grow in a well defined band, and platelike forms of *agaricia,* or lettuce coral, cascade down the reef face.

The shallow, sandy area behind Carysfort Reef is a beautiful sheltered sandy anchorage. Many varieties of coral, including this stinging fire coral, grow right up to the surface of the water, and can be explored by snorkelers. Photo: J. Halas.

Photographers hoping to capture portraits of Spanish grunt and other species should take care not to kick up the fine sediment at Carysfort with their fins. Photo: J. Halas.

At the base of the reef is a channel of fine sand, over 100 feet (30 meters) wide, that parallels the reef line. Across this sand channel a second offshore reef rises back up to 35 feet (11 meters) and then slopes down again on the far side.

Most boats anchor at a 40 foot (12 meter) depth or attach to the mooring buoy about 100 feet (30 meters) east of the tower. Plan to enter the water and swim east (offshore) to the reef face. Descending to the base of the reef at 65 feet (20 meters), you can begin your exploration in either direction. If there is a current, swim against it along the front of the reefline. When you have a thousand pounds of air left, make a surface check and work your way back to the boat.

This dive should probably be considered an intermediate dive simply because the depth is greater than 35 feet (11 meters). Inexperienced divers should be sure to monitor their time and depth and watch the current. If there are any photographers in the area, take care not to stir up the fine sand on the bottom.

The Gulfstream swings offshore as it passes Carysfort Reef, so the northerly current occurs less often and is not usually strong in this area. However, tidal currents moving offshore can be tricky, and it is important to realize the effects of the current and that it can change during a dive.

Near the light, divers may sometimes come across instruments, equipment, or markers associated with various reef research projects. Many of these are long-range studies that may take several years to complete. Although diving is not restricted here, scientists are hoping that people in the water will respect the importance of their work and not impair their efforts.

The Elbow 4

Typical depth range	:	12–35 feet (4–11 meters)
Typical current conditions	:	Variable (none to strong)
Expertise required	:	Novice
Access	:	Boat

When you look at a chart of the Upper Keys, it is easy to see that the Elbow is appropriately named. It juts out like a crooked arm as the reef dog legs to the north. The Elbow reef has been a catch-all for cargo ships, and its wrecks are the most notable feature of the area.

Veteran fish-feeders have been active at the Elbow for a long time. Hand-fed fish have been tamed by frequent feeding and plenty of attention. These silent creatures are responsive and certainly not camera shy. In fact, it almost seems as if they enjoy posing.

Diving Wrecks. There are three major wrecks on the Elbow and each makes an interesting dive. Furthest north and offshore is a metal-hulled wreck, probably a steamer, known as the *Towanda.* There is not much relief to the structure, which has been bent and scattered. Just offshore of the tower, the barge known by some as the *City of Washington* has settled into the sands and is now manned by a crew of lively fish. An older wreck, with wooden beams and iron fasteners or pins, lies just north of the tower. This is called the *Civil War Wreck*.

The remains of a wooden Civil War era wreck attracts a variety of fish at The Elbow, where the Key largo reefs curve to the southwest. Photo: J. Halas.

Cannons and anchors don't always mark the site of a wreck. When sailing ships ran aground, their crews often tried to refloat them by lightening the vessel. Heavy metal objects were normally the first things overboard. Photo: J. Halas.

Of these, the barge provides the most centralized dive spot and has the most to offer a diver with a full tank. This is also the site of most of the fish-feeding activity, so there are lots of friendly fish about. Photographers like this wreck because the standing sides and walls of jutting jagged metal make unique silhouettes, and folded plates form hideaways for shy species.

Other areas to examine range out from the barge. As you swim back to the end of the wreck nearest the light, a coral ridge flanks the barge. Over the ridge is a sand channel where larger groupers often wait until they can repossess their wreck from the divers.

At the other end, the coral shelf drops off into more sandy areas. In this direction, you may find the chain of giant links stretched across the reef. Also inshore along a sand channel, a bit of a distance away, is an old Spanish cannon left here long ago.

The Civil War Wreck is a place of isolated beauty. It sits in a flat area of sand and grass, so there is not much to explore around it. The wooden framework of the ship (held together with long iron pins) make a great boxy structure where all sorts of fish weave in an out.

Grecian Rocks* 5

Typical depth range : Shallow to 25 feet (8 meters)
Typical current conditions : Variable (none to moderate)
Expertise required : Novice
Access : Boat

Grecian Rocks is an exceptionally popular reef for snorkelers. The grass and sand on the back side provide good anchorage, and the shallow reef buffers the waves so the waters are very calm even on windy days. It is very easy for snorkelers, even beginners, to swim from the boat up to the reefline where the corals and brightly colored fish abound.

Corals. In the back-reef area, the elkhorn and staghorn colonies rise from the rubble zone along the fringe of the coral thickets. Here lazy barracuda keep a watchful eye on their visitors, moving slowly out of the way when swimmers come near. Large star coral heads cluster close to the surface, tempting swimmers to stand on their surface for rest and a break, however, anyone who cares about this fragile environment will not.

Grecian Rocks is a crescent-shaped patch. Hidden in the hollow of a large star coral in the middle of the back side of the reef is an old cannon, well camouflaged by its stony encrustation. At high tide, swimmers can venture into the shallower areas where the corals nearly touch the surface, but take care that wave surge is not great enough to thrust you into the rocky colonies.

The north end of Grecian Rocks makes an enjoyable, though shallow, dive for scuba enthusiasts. For the most part, the depth ranges from 15–20 feet (5–6 meters) around the end and toward the front of the reef. Many big coral heads form a bulwark that protects the shallower region of the reef.

Bright Spanish hogfish, tarpon and the occasional small reef shark can be seen at Grecian Rocks. Photo: J. Halas.

The area around Grecian Rocks has a number of old cannons and even a pile of cannonballs. Photo: J. Halas.

Cannons in the Park

Every once in a while, divers and swimmers unexpectedly come across a legacy from the venturous warships of long ago—isolated cannons on the ocean floor.

Two such cannons are located on what is known as the Cannon Patch, a small patch reef along the route to Grecian Rocks. Like most iron objects which have been underwater for a long time, these are heavily encrusted with a thick coral-like coating of calcareous oxide. They almost appear to be made of stone, but the elongated shape is clear as they lay juxtaposed on the bottom. Nearby are fused cannonballs piled up amid the waving gorgonians and busy fish that pick away at their surface. There is no record left of how these cannons came to be in this place; the solution of the mystery is left to our imaginations.

Observant snorkelers at Grecian Rocks can find another old cannon within the central hollow of a large star coral head in the middle part of the back reef. This small cannon, found elsewhere in the park, was placed there years ago by the park rangers. At the Elbow, sitting in a sandy channel aimed at nothing, is another large encrusted cannon.

French Reef 6

Typical depth range	:	Shallow to 45 feet (14 meters); 60–100 feet (18–30 meters) on deep reef
Typical current conditions	:	Variable, often strong
Expertise required	:	Novice to intermediate in depths less than 35 feet (11 meters); intermediate to advanced for deeper dives
Access	:	Boat

The outstanding feature of French Reef is its caves. No other dive site in this area has so many swim-throughs, overhanging ledges, and actual caves than this lovely reef located a mile and a half (two and a half kilometers) north of Molasses Reef. These are not the kind of caves that require special equipment. Rather, they are 3–4 foot (1 meter) high openings under the limestone rock that divers can peer into or swim through.

A stake marks the shallowest section. Extending from the shallows are limestone ledges separated by sandy swaths running out into deeper water. On top of the ledges great elkhorn colonies reach their arms toward the surface, and jumbles of staghorn branches form reef homes for the small tropical fish that dart in and out.

Blackbar soldierfish prefer dark areas, such as the caves at French Reef. Divers should be very cautious in entering any of the passageways through the reef, as many are dead-end tunnels. Photo: J. Halas.

A basket sponge and yellow tube sponge surmount a small hillock of coral at French Reef. Photo: S. Blount.

Marine Life. The corals grow in magnificent proliferation in the shallow water. Here, though, there seem to be more barren areas and the limestone rocks look older as the depth increases to about 30 or 35 feet (10 or 11 meters). It is in these areas that the clefts and breaks and overhangs have formed the extensive channels under the rocks that provide such exciting viewing.

Of course, with all these hiding places available, large fish are abundant on this reef. Divers often encounter huge grouper, and large green moray eels are a common sight. Here too, dog snapper are regularly seen swimming within the caves or along the ledges. This large snapper, uncommon at other reefs, is recognizable by the triangular patch under its eye. Smaller fish abound here also. The incessant yellowtail snappers surround anyone who brings food along, and usually a friendly French or gray angle will tag along for the dive.

Buoys. The secret to finding the exciting diving on French Reef is to be guided by the mooring buoys located on the offshore side of the reef. This was the initial test site for the unique mooring buoy system which is now found throughout the park/sanctuary area.

The southern stingray is fairly shy, and will flee from divers if given the chance. If necessary, it's well equipped to defend itself with the barb at the base of its tail. Photo: J. Halas.

Generally, the current will be flowing and usually the bow of the boat will face offshore to the southeast. If this is the case, enter the water and you will immediately see the entrance to Christmas Tree Cave just off the starboard side of the boat. This cave is named for the large conical star coral mound that rises over the top. This is a swim-through passage about 4 feet (1 meter) high with two entrances. The broadest opening is on the side toward the sandy bottom. On the other side, divers exit through an opening surrounded by corals. There is plenty of light and both openings can be seen at once. Pull yourself by hand along the 20 foot (6 meter) cave bottom. On the ceiling of the cave, bubbles of trapped air shimmer silver reflections of light. Red, yellow, and gray spongy growths coat the rock.

After exiting the cave, if the current is running in the opposite direction, swim to the Hourglass Cave next to the first buoy. This is about 150 feet (45 meters) in a southwesterly direction from Christmas Tree Cave. The best way to orient properly is to surface and take a compass bearing on the other buoy. After swimming over or around a coral ledge and across an open sandy area, the other cave should be visible.

As you venture inside this cave, it is readily apparent how it was named. In the middle is a vertical limestone column shaped like an hourglass that divides the cave into two sections. After exploring this area, follow the reciprocal compass bearing back to the first cave.

At this point, check your air. If you have about one thousand psi left, you can continue your excursion. From the wide sandy entrance to Christmas Tree Cave, swim directly offshore across the sand patch. Here, there is a limestone and coral "island" deeply undercut with crevices, recesses, and grottos on the north side. Though it is tempting to penetrate and explore these, be careful because some of them are dead-end passages and some are a tight squeeze for divers with tanks.

On the offshore end of this rocky formation, a large unnamed cave opens wide. It resembles the stage of a theater except the players are all fish which glide silently back and forth in stately display. This is as far as one needs to dive in this direction.

The third, fourth, and fifth buoys on French Reef center on White Sand Bottom cave, a large opening under a ledge that is home for many large fish. On the inshore side of this cave, a large sandy patch provides a good anchorage if the buoys are full. To the right, the third buoy is next to a rounded opening that breaches the reef wall. Offshore and to the left, the fifth buoy marks the entrance to a beautiful cave that is actually a crevice.

The two northernmost buoys mark a shallower area of ridges covered with elkhorn corals and numerous fire coral ledges. The south side of this area has beautiful star and brain coral heads. One of these has grown right over an old galleon anchor and the flukes stick out from under the coral.

Several large, tame groupers guard the entrance to Hourglass Cave at French Reef. Photo: J. Halas.

Molasses Reef 7

Typical depth range : Shallow to 40 feet (12 meters)
Typical current conditions : Strong northerly current
Expertise required : Novice to intermediate
Access : Boat

Molasses Reef is generally considered to be the most popular and most spectacular reef in the Upper Keys, and it lends itself to superlatives. Although it may have rivals for any one aspect, it really is deemed by many to have the most beautiful corals, the greatest relief, the most innumerable fish in the largest schools, and the clearest water.

This is a classic outer reef with a well-defined spur and groove system of coral development. The shallow coral ridges begin near the lighthouse and extend out toward the deeper water. At a depth of about 35 feet (11 meters) the massive corals terminate and a zone of low profile relief extends offshore to deeper water.

Mooring Buoys. Because of the severe diving pressure here, there was a serious problem with anchor damage to the corals. Now the sanctuary administration has installed a series of mooring buoys, large blue and white plastic floats, to which boats can attach and not have to anchor. Unlike former buoy styles, these do not have large heavy chains and concrete bases (which cause as many problems as they solve), but rather are drilled directly into the bottom and installed. The buoys are available to everyone on a first-come first-served basis.

A ship's winch marks one of the outstanding dive areas of Molasses Reef, a large sandy basin surrounded by coral heads. Photo: J. Halas.

Keys nightlife below water is as colorful as it is above, with nocturnal animals such as this scrawled filefish cruising the reefs after dark. Photo: J. Halas.

The northern sector of Molasses Reef has no mooring buoys, although the reef development there is every bit as spectacular as the rest of the area. The reason is an unspoken rule that discourages diving in this portion of the reef to allow for safe access by the *MV Discovery*, the big blue glass-bottom boat operated by the Pennekamp Park concessionaire.

The Winch. The central portion of Molasses, squarely offshore of the lighthouse, is one of the most beautiful dives along this reef. It is commonly called the Winch and a dive here usually begins close by this feature. Occasional wreckage, an indication of the many ships that have met their end on these rocks, is scattered all around this area, but the most dramatic evidence of a ship's desolation is a lone winch or windlass that sits isolated in the center of a barren sandy patch near the middle of the reef, in about 30 feet (10 meters) of water.

With the winch as a landmark, it is quite easy to describe a dive in this area. You'll want to explore the perimeters of the basin that the winch sits in. To the east, just about opposite the lighthouse, a sandy channel cuts between towering coral walls and leads seaward. On the left, astounding numbers of fish huddle under and around the corals that make up the ravine wall. A maze of massive star coral heads is just seaward of this area. Look closely at these huge cone-like shapes and you will see that they sport their own decorations. Many colorful Christmas tree worms that grow in vivid spirals and look like tiny flowers sprout from the coral colonies. When any moving thing comes too close, they zip back into their holes and only reappear when their fear is gone.

White Bank Dry Rocks* 8

Typical depth range : Shallow to 25 feet (8 meters)
Typical current conditions : None to moderate
Expertise required : Novice
Access : Boat

Boats on the way to French Reef must take care as they pass between two shallow patches known as White Bank Dry Rocks South and North. White Bank is an extensive sand bank that stretches north and south along the offshore reaches of Hawk Channel. Along its edge are numerous patch reefs and isolated coral heads which make excellent sites for snorkeling or fishing. The White Bank Dry Rocks are toward the south end of the park/sanctuary and make up the largest and loveliest of this area's sea gardens.

Both of these twin patches are fine snorkeling spots. Behind the reef is a good anchorage with calm water and the reefs themselves are prime examples of outer patch reef development with staghorn and elkhorn beds around the fringes. Small star and brain coral heads are scattered over the reef flat, and many spectacular species of soft corals produce a true garden effect.

Large elkhorn patches in shallow water make White Bank Dry Rocks appealing to divers and snorkelers. Photo: J. Halas.

A common hogfish or hog snapper takes a nocturnal rest at the base of a gorgonian and sea fan. The sea fans at White Bank are above the surface of the water at low tide. Photo: J. Halas.

Aquarium in the Keys. These patch reef areas commonly house the small tropical fish that collectors love to display in their aquariums. Don't forget, though, that in this protected sanctuary collecting of any sort is not allowed, in order to ensure that these beautiful areas will be a legacy to the future. However, do enjoy the experience that comes closest to swimming in an aquarium filled with these gorgeous tropicals. You will see many juvenile species here. Most are quite territorial and stay pretty close to their own special habitat. The small French angel, basically black, has distinctive vertical yellow stripes. Small blue and gold striped angels may grow up to be either queen or blue angels—their juvenile forms are very similar. Golden rock beauties have a black spot on their back sides which grows as they do. By the time these fish are adults, the black area has grown many times its original size.

One spot on White Bank Dry Rocks South is especially good for a mixed trip of snorkelers and divers. At the far south end of this reef, the edge of the patch drops off into a grassy gully about 15 feet (5 meters) deep. From here snorkelers can swim up onto the shallower reef flat, and divers can investigate the corals along the edge and front side, which reaches a depth of about 25 feet (8 meters).

Off to the right, about 40 feet (12 meters) south of the main patch reef area, is an isolated coral ridge made up of elkhorn colonies and star coral heads. Rising almost 10 feet (3 meters) from the surrounding bottom, this relatively small area is easy to explore and perfect for macro or close-up photography and available light pictures.

Pickles Reef 9

Typical depth range	:	10–25 feet (3–8 meters)
Typical current conditions	:	Variable but moderate
Expertise required	:	Novice to intermediate
Access	:	Boat

Pickles Reef is a versatile diving area located a little more than 2 miles (3 kilometers) southwest of Molasses Reef. The origin of its name is obscure, but scattered around a small area just seaward of the stakes marking the reef are a number of what appear to be hardened encrusted pickle barrels. Actually, these small kegs are probably containers of mortar mix or cement that were destined for the great brick forts of Key West and the Dry Tortugas but were lost here.

Like the other outer reefs, Pickles Reef has collected its share of wrecked vessels. The most recent, a large shrimp boat, was stranded on the coral crest in about 10 feet (3 meters) of water, and its remains still rise from the water at low tide. The huge engine, parts of the hull, and the remaining netting and rigging have become a home for many of the smaller species of fish which are almost immediately attracted to any new habitat that is suddenly added to the reef environment.

Plants and Shells. Common on this reef are deep purple sea fans tenaciously anchored to the bottom by the strong grasp of their stems and resolutely facing the wave surge toward the southeast. These lacy fan-like

Spectacular mollusks, such as the queen conch and the flamingo tongue shell, can be found in profusion at Pickles Reef. Conchs are protected by state law, and no one may take them for any purpose other than food. Photo: J. Halas.

A schoolmaster poses in front of a sea fan at Pickles Reef. Sea fans, many with beautiful flamingo tongue shells on them, are common on Pickles Reef. Photo: J. Halas.

forms are far less delicate than they appear as they sway gently back and forth. They are the province of one of the most intriguing photographic subjects for close-up or macro photography, the flamingo tongue shell which commonly grazes the sea fan for algae and debris trapped in its filigree network. The flamingo tongue, relatively common at Pickles Reef but not always found in similar locations, is about an inch (25 millimeters) long and has an oval flesh-colored shell. Its glory is the magnificent animal within. The mantle, or skin, which emerges and covers the highly polished shell, is a transparent yellow with golden leopard spots outlined in black.

Another shell found in the rubble area and grass flats on the back side of Pickles Reef is the queen conch. This is a good area for collecting these large mollusks, which graze here for their food. Although well camouflaged as nondescript rocks, conchs are easy to spot and retrieve once you know what to look for.

Queen conchs are protected by Florida law. They may not be taken for any purpose other than for food, and no one may possess more than ten conchs at one time. Although there is no official size limit, smaller conches without a flared lip (known as rollers) should be left to mature.

Anchoring offshore of the stakes at Pickles Reef in about 25 feet (8 meters) of water provides a good central location for exploring this interesting reef with its well-developed spur and groove system. Although the front side has less relief and lower corals than some of the other reefs, they are no less beautiful.

Conch Reef 10

Typical depth range : Shallow to 100 feet (30 meters)
Typical current conditions : Variable (moderate to strong)
Expertise required : Novice to advanced
Access : Boat

Conch Reef is an extended shallow area that stretches for over a mile (one and a half kilometers) along the outer reef line. It is marked with a red nun buoy located toward the south end of the main section. There is a steep drop from 60 feet (18 meters) to about 100 feet (30 meters) known as the Conch Wall just offshore of the red can. South of there, an unusual trough or tongue 40 feet (12 meters) deep cuts in through the reef. All along Conch Reef are numerous sites where exciting diving is commonplace.

Conch Reef is one of the areas where coral harvesting nearly devastated the reef, finally prompting state legislation which prevented the taking of corals. Now all species of corals are protected, and possession and sale of any Atlantic or Caribbean species is illegal. On the ledge that runs along the top of the reef, the stumps of pillar coral colonies are still visible. This rare and protected coral has not completely died here, however, and new colonies have begun to reproduce and grow.

One such area, the Pillar Patch, is an interesting site for a shallow dive or snorkel. It is located about one half mile (one kilometer) along the reef north of the red can on the ledge, about 15 feet (5 meters) deep. Inshore, this rocky ridge drops off into sand at a depth of 18–20 feet (6–7 meters).

Along the rim of the dropoff at Conch Reef, the flow of water promotes the growth of filter-feeding animals such as this deepwater gorgonian, a soft relative of the coral. Photo: J. Halas.

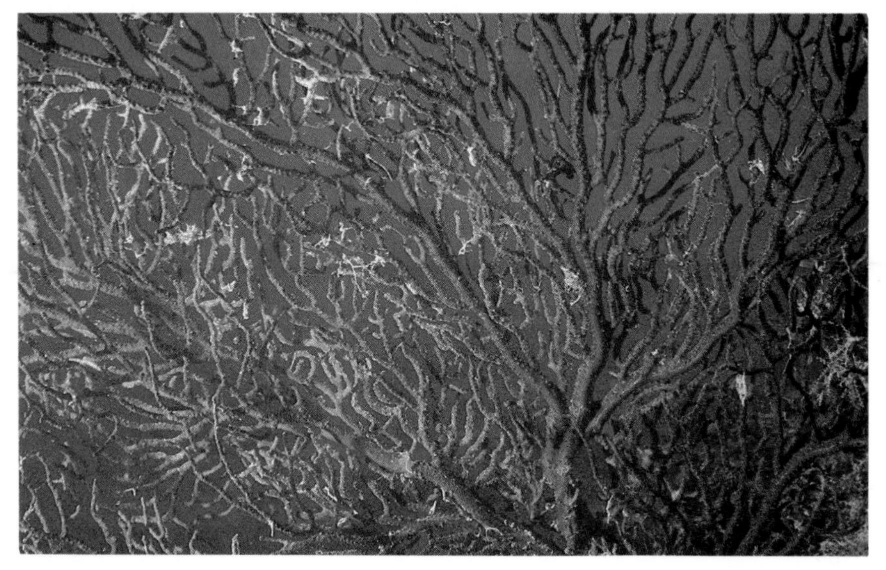

Basket sponges are also common along the dropoff here. The wealth of tiny organisms rushing by in the current causes the sponges to grow to even more spectacular dimensions. Photo: J. Halas.

Marine Life. The rim here does not follow a straight line, but instead "scallops" in and out with the sheerest drops occurring along the promontories or bluffs that project seaward. On these points, which get the full force of the currents, sea life grows in a variety of forms. Lacy deepwater gorgonians spread out perpendicular to the current like giant filigree fans along the top of the bluff, and huge purple sponges stand like giant empty vases. Everywhere small species of deepwater tropical fish dart back and forth investigating the bottom growth.

The dive charters that bring divers here usually anchor along the rim of the drop in 60 feet (18 meters) of water. You can enter the water and swim to the anchor line to make your descent. Immediately apparent from the surface are a number of sand channels cutting through the coral rock and looking like rivulets of sand running down the slope. On the bottom, follow one of these grooves down to your desired depth. Working upcurrent along the base of the reefline, be sure to monitor air, time, and depth gauges regularly. Midway through the dive, ascend a bit up the slope and work back to the anchor line.

Hen and Chickens 11

Typical depth range	:	20–22 feet (6–7 meters)
Typical current conditions	:	Usually none
Expertise required	:	Novice
Access	:	Boat

Hen and Chickens is a rather unusual diving area for an inshore patch reef because of the depth and the size of the coral heads. Like a bunch of little chicks clustered around a mother hen, this reef is a collection of large star coral heads rising steeply from the bottom and marked by a navigation light. Located less than 3 miles (5 kilometers) from shore, it provides easy access for smaller boats and for divers who have a limited amount of time, yet want a good dive with plenty to see.

Damaged coral. Hen and Chickens received serious scientific attention when a number of the large star coral heads suddenly died in 1970. Almost 80% of the reef was killed after a particularly cold winter when the corals were subjected to lower than average water temperatures for an extended period. The cold waters, laden with sediment stirred by the winter winds, flowed out of shallow Florida Bay in the Gulf, through the opening between the Keys at Snake Creek, and overwhelmed this beautiful assemblage of massive corals.

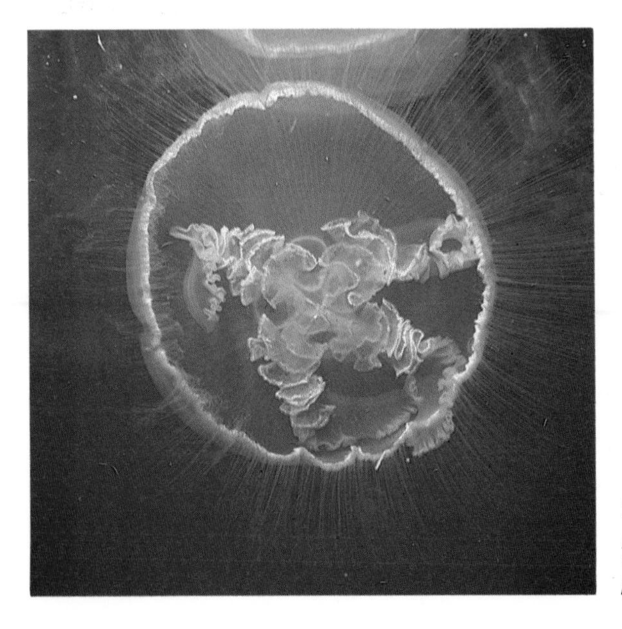

The underside of a moon jellyfish shows its trailing veil of stinging tentacles. Photo: J. Halas.

The jacknife was named for its resemblance to a half-open pocket knife. The tiny tropicals are popular with salt water aquarium enthusiasts. Photo: J. Halas.

Today, although much of the living coral is gone, the coral structures remain. Their surfaces have been repopulated with soft corals, and now this secondary growth of waving plumes, branching candelabra forms, and broad sea fans adds graceful beauty to the rock contours. Large fish still hide out in hollows beneath the heads, and grouper, grunts, spadefish, sheepshead, porkfish, and snook are commonly seen.

Spearfishing is not allowed here because this reef is still within Florida waters. In the Upper Keys, as far south as Long Key Bridge, spearfishing is only permitted beyond the 3 mile (4.8 kilometer) limit. The snook which cruise under and between the coral heads here are doubly protected. This elongated silver fish, looking somewhat like a long-snouted barracuda with a black stripe, is prized by seafood connoisseurs for its delicious flesh. However, it may not be taken with a spear or gig. Fishermen are limited to a catch of two snook which must be over 18 inches (46 centimeters) long.

At Hen and Chickens, the main concentration of coral heads is a bit inshore and to the north of the light. Boats should anchor so divers can best cover this area. Remnants of modern wreckage lie just inside the coral clumps on the shoreward edge of the reef. Occasional lobster can be found amid the general debris under the light, and large barracuda patrol the area here.

As with other inshore areas, the visibility can vary greatly. Local weather will usually determine the quality of the water clarity.

Keys Bridges 12

Typical depth range	:	8–20 feet (3–6 meters); average depth is about 12 feet (4 meters)
Typical current conditions	:	Variable and continually changing; very strong currents encountered
Expertise required	:	Intermediate to advanced (should be accompanied by someone experienced with these conditions)
Access	:	Shore or boat

Unfortunately, the most accessible diving in the Keys is not suitable for all divers. The Keys bridges are, in effect, mini-reef areas where corals, sponges, fish, and lobster are found in abundance. For the novice diver, however, it is an area of unsuspected dangers from the swift changeable currents encountered here.

Because of the tidal flow, great amounts of water are being funneled through a narrow constricted passage between these islands. The nature of the current is constantly changing as the tide rises and falls, and this must be taken into consideration when planning a bridge dive. The best time to begin a dive is the period just before the slack tide when the current is decreasing. The rush of the running water stops altogether, generally at mid-tide, when the direction of the flow is switching. The length of time for the slack varies and usually depends on factors such as the wind, the height of the tide, and the phase of the moon.

After the direction of the current changes, the force of the flow increases slowly and a dive is manageable for an hour or so. A scouting mission to the bridge the day before a dive is probably a good idea so that the amount of current to contend with can be assessed and the times when the current is slack, or flowing less strongly, can be determined. Remember, however, that the slack (like the tide) occurs about 55 minutes later each day.

Corals which ordinarily form round boulders or heads, such as this brain coral, are found flattened and spread out under the various Keys bridges. This growth pattern may be due to the volume and velocity of water that rushes through the narrow channels during tide changes. Photo: J. Halas.

Diving the Keys bridges requires preplanning and careful follow through of the dive plan. Those who take the time will be rewarded with a view of a unique ecosystem and abundant marine life.

Corals. The currents and the generally murky water have had unusual effects on the ecology of the area under the bridges. Life has adjusted to the flow, which reverses every six hours. Generally corals would not be found this close to shore, but they grow here because of the tremendous energy generated by the tidal flow and the amount of food carried by the water. However, the forms have adapted to the conditions. Some corals which usually grow into rounded coral heads, like star and brain corals, have flattened and spread out over the bottom so that they will receive the greatest amount of light and exposure to the current. Likewise, an encrusting form of orange sponge is common and coats rocks and bottom with splashes of brilliant color.

Fire coral especially responds to the high energy and the availability of food and grows amazingly fast, coating almost everything. Also nestled into every crevice and hole and under every rock are myriad black long-spined sea urchins.

The most commonly dived bridges include the Snake Creek and Whale Harbor spans in the Upper Keys. These are not really recommended because of their narrow channels and the number of boats that pass through this area. Below Islamorada, the four bridges at Indian Key Fill are popular, as are the two south of Lower Matecumbe, Channel #2 and Channel #5. Channel #2 probably has the greatest area of hard bottom extending outward from the bridge, while Channel #5 has grassy bottom along the marginal edges.

Long Key Bridge is different because of its great length; it is the second longest of the Keys bridges and is characterized by numerous large slabs beneath the footers. South of Marathon, the old Seven Mile Bridge—the Keys' longest bridge—is also much shallower in places than the others. Both Bahia Honda bridges are excellent dives, with deeper depths and hard bottoms extending outward from the structure.

El Infante and the Spanish Treasure Ships 13

Typical depth range	:	12–15 feet (4–5 meters)
Typical current conditions	:	None to moderate
Expertise required	:	Novice
Access	:	Boat

In 1733, a fleet of Spanish ships had left Havana and were heading northward, laden with New World riches destined for the king of Spain. When they left the coast of Los Martires, now called the Florida Keys, they were driven onto the reefs and into the shallows by a vicious hurricane. The remains of these vessels lay camouflaged by the passing years, until modern technology allowed us to search beneath the sea for its hidden treasures. The clue to the location of these ships is an innocuous pile of rocks on the ocean floor.

The *Infante* was such a ship. Nearly an acre of ballast stone, ranging in size from small fist-sized cobbles to huge boulders, was strewn over the shallow bottom, marking this treasure wreck with probably the largest ballast pile in the Keys. In the middle of the site are large timbers protruding from the thickest part of the scattered rocks. These survived, protected from the teredo worms by the sand until recently, and are all that remain of the planking of the ship's deck.

As with many of the early wooden wrecks in the Keys, a pile of ballast stone is all that remains of the El Infante. *The rocks were carried in the lower hold of ships to lower their center of gravity and keep them stable in heavy seas. Photo: J. Halas.*

A scorpion fish rests near the site of the Infante. *These bottom-dwellers look like rocks, sometimes hiding among piles of ballast stone. Spines along their dorsal fins can inject a powerful poison if they are stepped on or brushed. Photo: J. Halas.*

Buried Treasure. There is probably no need to remind anyone to look carefully for objects that may be of value. If you uncover something made of gold, you will know it immediately; this precious metal is not affected by long years underwater and gold objects appear much as they did when they were lost. A friend wears a carved floral wedding band made of gold that he found deep in the sand on this wreck. Silver is another matter. The saltwater causes silver to form a crust of oxidation and the metal blackens. Many silver coins look more like rocks than money, but watch out for the odd shapes and the weight of the coin. Of course, a metal detector helps distinguish them immediately. This wreck is famous for the round pillar dollars, beautiful and valuable coins that were first minted by the Spanish.

Most people are thrilled with finding anything that has not been handled or exposed for over two hundred and fifty years. Items like this are pieces of glass from old hand-blown bottles, pottery fragments, nails and pins that were used to build the ship, armament, and even the ballast stones themselves.

Before abandoning the search on *El Infante*, be sure to take a look around the perimeter of the wreck and see this pretty area of Little Conch reef. Offshore, fire coral ledges and isolated heads provide habitat for fish and lobster. To the south and inshore a bit are more ledges and an excellent chance to pick up your dinner.

4

Diving the Middle Keys

Middle Keys. Long Key Bridge essentially marks the transition point between the Upper and Middle Keys. South of here are the crowded commercial fishing community of Conch Key, the exclusive residential community of Duck Key, and the relatively uninhabited expanse of Grassy Key.

The center of Marathon, "the Heart of the Florida Keys," is at milemarker 50. This community is the second largest in the Keys after Key West, and it features a full-sized air strip with daily commercial service from Miami. Many businesses are established here and the full range of services are available for the tourist. Fisherman's Hospital is a fully staffed community hospital well equipped to care for almost all serious medical problems and emergencies. The Highway Patrol and Sheriff's stations are located here, as are the Middle Keys courthouse, high school, library, and other county agencies.

Beyond Marathon lies the wide expanse of ocean spanned by the Seven Mile Bridge. The newly built structure is only slightly under seven miles (11 kilometers) long and is paralleled by the original bridge, which provides the University of Miami field station at Pigeon Key with access to the highway. The upright posts and railings of the old bridge rate a careful look; they are made from the original rails of the railroad. From the vantage of the new bridge it is easy to see the arches and footers of the bridge that became known as one of the wonders of the world. Off to the left, Sombrero Light stands its vigil more than five miles (8 kilometers) out to sea.

Three small keys, one devoted entirely to campers, lie between the Seven Mile Bridge and Bahia Honda Key. Bahia Honda State Park is located here and is well known for its swimming and camping facilities. This is also the headquarters for the Looe Key National Marine Sanctuary. A glance at the old Bahia Honda Bridge (visible on the left from the new bridge) leaves no doubt as to its original purpose as a railroad trestle. The iron structure was too narrow to use as a roadbed, so the pavement was laid on top of the bridge!

It seems the style of the Keys becomes progressively more casual the farther south one travels. The focus of the hotels and resorts of the middle Keys is on sun, sand and relaxation, and divers are welcomed. Photo: S. Blount. ▶

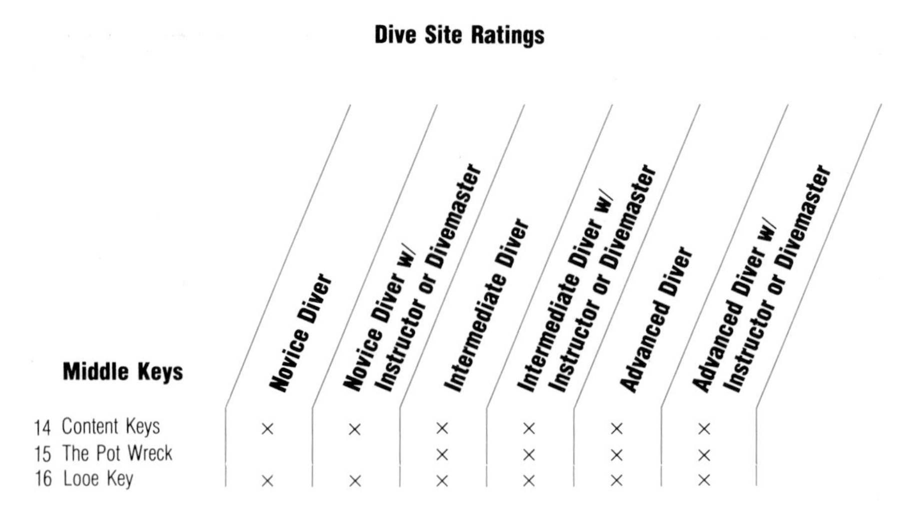

Dive Site Ratings

Middle Keys	Novice Diver	Novice Diver w/ Instructor or Divemaster	Intermediate Diver	Intermediate Diver w/ Instructor or Divemaster	Advanced Diver	Advanced Diver w/ Instructor or Divemaster
14 Content Keys	×	×	×	×	×	×
15 The Pot Wreck			×	×	×	×
16 Looe Key	×	×	×	×	×	×

When using the accompanying chart see the information on page 4 for an explanation of the diver rating system and site locations.

Diving in the Middle Keys

Though only three sites are covered in the Middle Keys, there are many excellent sites that can easily be explored, and should not be missed. Some of these sites are: Sombrero Reef, Delta Shoals, Seven Mile Bridge Reef, and the bridges in the Keys.

Many of the Keys' bridges have fishing catwalks which are very popular. Be extremely careful when passing through these areas to avoid the trailing hooks and sinkers which are constantly being thrown in and retrieved. There is always quite a bit of monofilament and leader wire wrapped around the bottom features and it is easy to get tangled up or snagged in this.

Lobsters and Fish. Most people dive the bridges to search lobster since these creatures are numerous here and find plenty of habitat to hide in.

Large fish are often found beneath these bridges, and probably the most magnificent are the huge silvery tarpon which are commonly seen. For unknown reasons, they tend to gather in certain preferred areas, and divers who encounter them will enjoy the experience of swimming among these monstrous fish that often weigh over 100 pounds (220 kilograms). Large sharks pass through these channels but are rarely seen during the day. Sometimes huge turtles swim by on their way from one side of the Keys to the other. Barracuda are common, as are grouper, sheepshead, mutton and mangrove snapper, and many large French, blue, and queen angel fish.

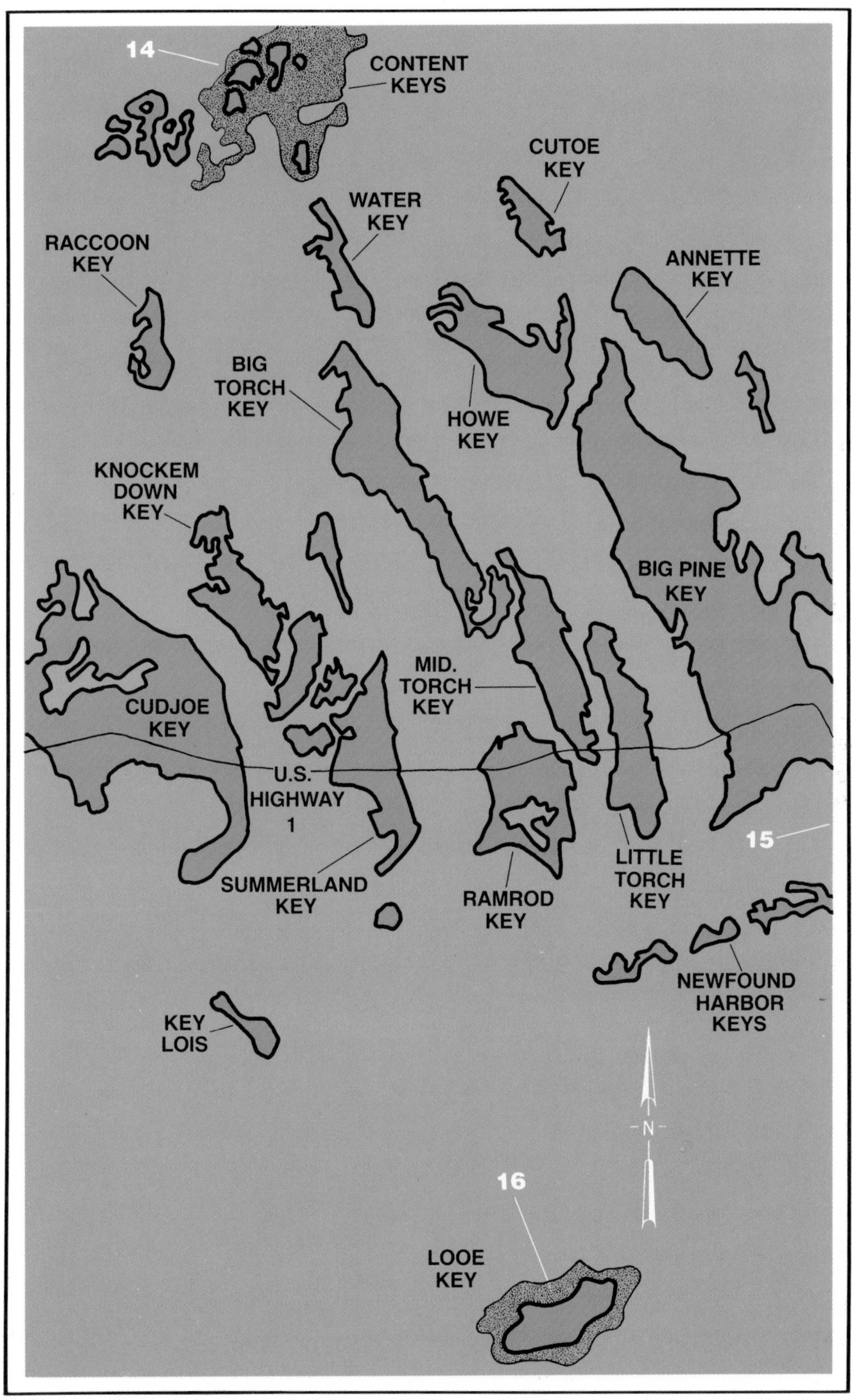

From Islamorada to Big Pine Key, the middle Keys offers fine restaurants, tropical scenery and plentiful diving services.

Content Keys 14

Typical depth range	:	8–15 feet (1–4.5 meters)
Typical current conditions	:	Minimal
Expertise required	:	Novice
Access	:	Boat

The Content Keys, are uniquely positioned on the Gulf side southwest of Marathon. Sheltered by the low-lying string of islands nearby, this location is in the lee of the prevailing southeast wind and remains quite calm even on the windiest days. This is not the case if the wind is coming from the west or the north, however, when this occurs the ocean side is often sheltered. Although not a preferred dive site when the clear Atlantic reefs beckon, the Content Keys are known to conjure up their own special magic.

Rounded brain and star corals can be found in the area of the Content Keys, a sheltered Gulf side dive site. When the winds blow hard from the south or east, the Content Keys remain diveable even though the Atlantic side reefs are blown out. Photo: D. Kincaid.

Two crustacean delicacies are found in abundance near the Content Keys, lobsters and stone crabs. Though both are easy pickings for divers, check the game regulations before stuffing any in your game bag. Photo: D. Kincaid.

Marine Life. Here is the chance to experience a really productive dive for lobster and stone crab, as well as participate in a unique ecological environment. The Gulf side of the Keys supports an entirely different eco-system from the ocean side. Occasionally, corals which tolerate greater changes in temperature and lower light levels grow here. These include the rounded brain and smooth starlet corals. Fish common to this area are sheepshead, redfish, trout, red grouper, jewfish, and various snapper species as well as an outstanding variety of juvenile tropicals that seek refuge in the features of this reef area. Dives usually begin at the Content Ledges just offshore from the Keys. This rocky shelf extends east and west and rises 2–3 feet (.5–1 meter) from the 8–10 foot (1–3 meter) sea bottom. Numerous pot holes and cracks in the coral rock make an ideal habitat for both lobsters and stone crabs. Even the locals come to this area for these delicacies.

Offshore, about a ten minute boat ride away, a collection of large coral heads, rare on the Gulf side, rise from a depth of about twelve feet (4 meters). Here divers can spread out and wander among these massive living boulders.

Realistically, this dive site is shallow and the visibility rarely reaches the "crystal" quality, but when weather is bad, and most boats can't leave the dock, it is nice to have a place to go diving.

The Pot Wreck 15

Typical depth range	:	90–105 feet (27–31 meters)
Typical current conditions	:	Variable; moderate to strong
Expertise required	:	Intermediate to advanced
Access	:	Boat

Sitting upright on the bottom with a definite list to the port side, the "Pot Wreck" or "Cannabis Cruiser", is a good dive for a small group of experienced divers. A former trawler, this 65 foot (19 meter) steel vessel was apparently about to be discovered with its illegal haul and was scuttled on purpose. A dive here will not yield contraband, but does make an exciting excursion to the bottom of the sea.

Preparation for a dive to this depth is essential. Divers should be well-equipped, experienced, and comfortable in open water. Because of the current, a trailing line with a float extending behind the boat is an important safety measure. No diver should descend without plenty of air, a pressure gauge, buoyancy compensation, a watch, and his dive tables at the very minimum.

Professional dive charter boats dive the pot wreck regularly. The trick to a good dive here is placing the anchor in the proper location. The anchor should be either on, next to, or slightly down current from the wreck so divers do not have to waste precious bottom time swimming to the wreck.

A sponge clings tenaciously to the side of the "Pot Wreck." Its holds full of marijuana, this steel vessel was apparently scuttled, finding bottom in over 100 feet (30 meters) of water. Photo: D. Kincaid.

Various encrusting organisms find the "Pot Wreck" a perfect home. The hull is covered with spiny oysters. Photo: D. Kincaid.

The twenty minutes allotted to divers at this depth is enough to get a good look at the wreck. The superstructure consists of a large cabin. What may once have been a flying bridge or lookout stands above, while a railing runs forward around the bow and a large flat deck area extends out above the stern. All but one window in the cabin has been broken and there are three open doorways so entry into this area is easy. Gone now are the steering wheel and equipment from the command station, but some clutter remains. An open forward hatchway beckons to those divers equipped with lights. Interior exploration should be pre-planned and pre-arranged with the divemaster. The holds are packed to capacity with bales of marijuana, now totally ruined by immersion in the sea, but still readily identifiable. Most divers cannot resist the temptation to reach in and feel the slimy weed which disintegrates immediately and drifts off in clouds carried by the current.

Looe Key Reef 16

Typical depth range : 5–35 feet (2–11 meters)
Typical current conditions : Light to moderate
Expertise required : Novice to intermediate and advanced
Access : Boat

About 7 miles (11 kilometers) south of Ramrod Key lies one of the loveliest, most varied and prolific reefs in the Florida Keys, Looe Key.

In 1744 Captain Ashby Utting accidently ran his frigate, the *H.M.S. Looe*, hard aground, thus giving the key its name. The remains of the ship lie between two fingers of coral near the eastern end of the reef about 200 yards (185 meters) from the marker. Although the ballast and a heavily encrusted anchor remain, only the trained eye can distinguish them from the reef. Since all of Looe Key is now designated a marine sanctuary no spear fishing, lobstering, or artifact recovery is permitted. "Look but don't touch."

Looe Key is a geological anomaly. It is totally unlike any of the reefs of the lower keys in structure or in variety of life. Looe Reef contains species of coral found in both the patch reefs and outside reefs as opposed to other sites further west that only contain fingers of fire coral, elkhorn and staghorn coral.

Looe Key, a national marine sanctuary, is a perfect example of a spur-and-groove reef system. Long fingers of coral fan out over a white sand bottom, which forms gullies between the rows of coral heads. Photo: D. Kincaid.

Coral polyps and feather worms form a modern-art mosaic at Looe Key, named for the HMS Looe, *a British ship which sank here. Photo: D. Kincaid.*

Looe also has large quantities of sand at its base, which spill out slowly in a never-ending cascade southward into the ocean abyss nearby, and is washed constantly by the waters of the Gulfstream. Looe Key's pulsating life cycle is continually renewed by larval creatures that drop from the sargasso weed and propagate the reef. Although parts of the reef are awash, the fingers of coral base have gullies approaching 35 feet (11 meters) deep in some spots, topped by monastera and waving sea fans. This unique structure, blue water and moderate current combine to make an idyllic dive.

Marine Life. Temperatures year round seldom require more than a wet suit top. Fishes from several different environmental zones congregate here—burrowing yellow headed jawfish, parrots, and surgeon fish on the north side behind the reef rubble. Barracudas and jacks on top of the reef, grunts, cotton wicks and angels on the reef proper, and rays, lizard fish and an occasional peacock flounder on the sandy floor on seaward side. Several commercial dive boats visit Looe Key; most leave from Big Pine Key.

The reef is roughly 200 yards (185 meters) wide and 800 yards (750 meters) long and is roughly Y-shaped. In short, Looe Key is a rewarding and exciting dive for all levels of expertise and is probably the most beautiful reef in the lower keys.

5

Diving the Lower Keys

Divers from the frozen north who drive into the Lower Keys enter a wonderland of wet delights. As the only highway leaps from bridge to bridge, just a short swim or boat ride away are reefs, shallows, islands, and currents unparalleled in the diversity of their wildlife. This area has been described as "a land that's mostly water and a sea that's mostly sky."

Geology. The Keys island chain runs roughly east and west, not north and south as many people assume. On the north or Gulf of Mexico side are dozens of mangrove islands and wading birds. On the south or Atlantic side are the reefs, washed by the Gulfstream—that great river in the sea that meanders up the east coast of the United States.

The Lower Keys contain several basic geological and environmental zones. Southernmost are the Outside and Barrier reefs, with fishes and coral influenced by the Caribbean and Gulfstream; immediately to the north, between the reefs and the Keys, Hawk Channel has a generally muddy or sandy bottom and intermittent patch reefs that rise very near the surface in some areas. Further north are the Keys proper, or back country, a mangrove nursery area for most of Florida's commercial fisheries and essentially an estuarine environment similar to and influenced by the Shallows of Florida Bay, the Everglades, and finally, the Gulf of Mexico, which is a large, relatively shallow partially enclosed sea. Because of this mix of marine influences, the Keys generally have a wider variety of fish and coral than even some areas of the Caribbean. The transitional area between the Keys and the Gulf is a hard rocky ledge that runs from the Content Keys north of Big Pines and southwest past Cottrel Key. This ledge is similar to what is called Ironshore throughout the Caribbean, a sort of compact conglomerate of shells, coral, and fine sand fused together by thousands of years of pressure. Occasionally, this rocky substrate will be cratered like Swiss cheese, where pockets of organic material have rotted and formed an acidic solution that eats huge holes called solution holes. Many of these holes are aggravated and enlarged by wave action.

▶ *Divers examine an unusual coral formation at Sand Key, in the lower Keys. Photo: D. Kincaid.*

DIVE SITE RATINGS

LOWER KEYS	Novice Diver	Novice Diver w/ instructor or Divemaster	Intermediate Diver	Intermediate Diver w/ instructor or Divemaster	Advanced Diver	Advanced Diver w/ instructor or Divemaster
17 South Beach Patch	X	X	X	X	X	X
Key West Harbor					X	X
18 Sand Key	X	X	X	X	X	X
19 Outside Reefs			X	X	X	X
20 Rock Key and Eastern Dry Rocks	X	X	X	X	X	X
21 Western Dry Rocks (The K Marker)	X	X	X	X	X	X
22 Chet's Wreck #1	X	X	X	X	X	X
23 Chet's Wreck #2 (The Aquanaut)			X	X	X	X
24 Cottrell Key (Gulf Side Reef)*	X	X	X	X	X	X
25 The Lakes*	X	X	X	X	X	X
26 Marquesas Keys	X	X	X	X	X	X
27 The Wrecks in Marquesas	X	X	X	X	X	X
28 Cosgrove Shoal			X	X	X	X
29 Marquesas Rock			X	X	X	X

*Indicates Good Snorkeling Spot.

When using the accompanying chart see the information on page 4 for an explanation of the diver rating system and site locations

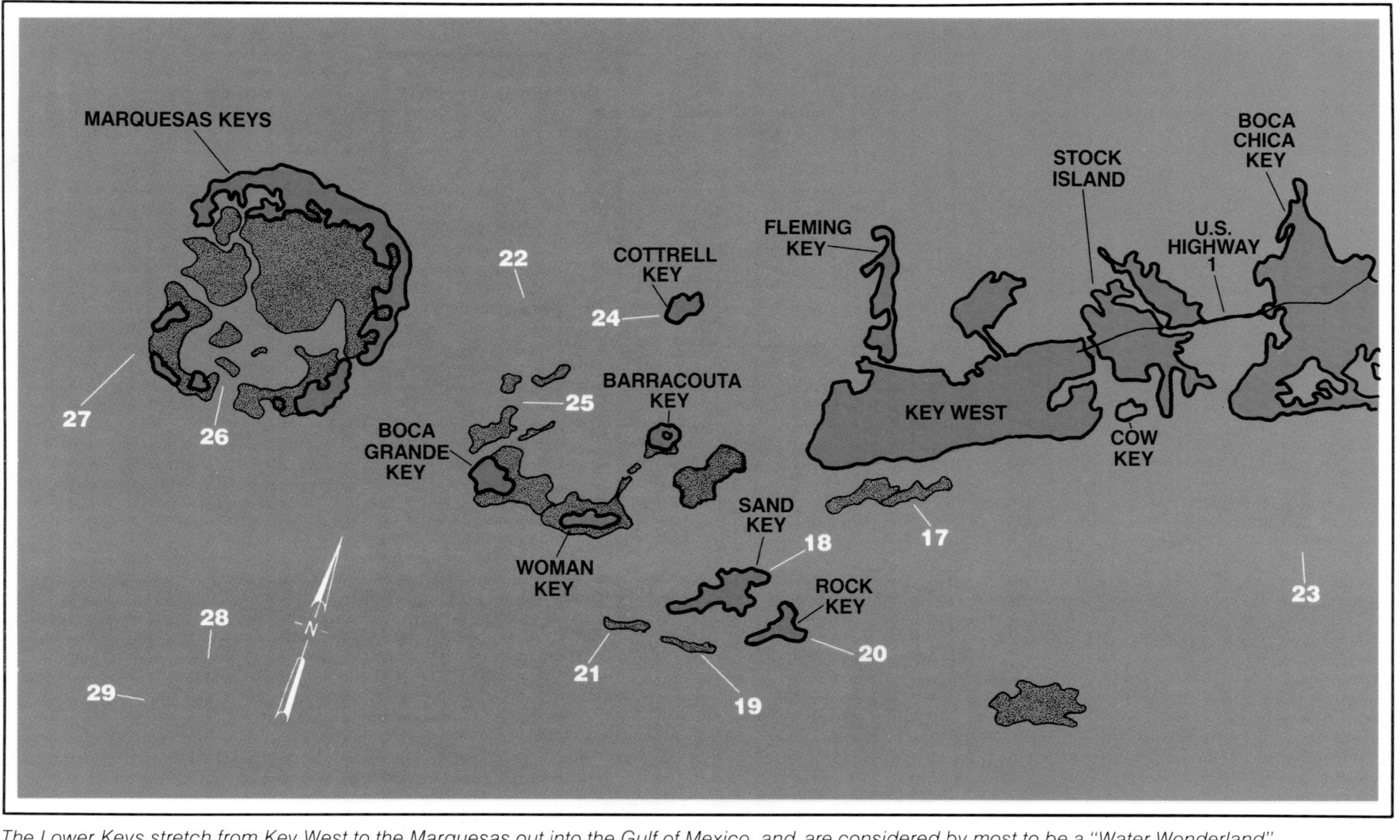

The Lower Keys stretch from Key West to the Marquesas out into the Gulf of Mexico, and are considered by most to be a "Water Wonderland".

Unique Diving Services. Most of the dive shops in the Keys offer some type of charter dive service, but there are several, in my opinion, that are unique.

Earl DeTurk of Big Pine Key offers custom dive trips aboard his 41 foot (12 meter) Chinese Junk-rigged trimaran *Water Spyder.* Earl is a naturalist and consequently looks at the Keys through different eyes than most of us. He not only day sails *Water Spyder* but will do week-long excursions on request. If you hate sailing on conventional keel boats, *Water Spyder* is a treat because it sails flat and doesn't roll at all. You can actually set your beer down in a civilized manner instead of sailing on your ear. Earl does not do scuba trips, only free diving from this unique craft. Earl works out of the Looe Key Reef Resort.

Another special trip is given by Captain Vicki Impallomeni. This vivacious lady is also a marine biologist, and she specializes in family trips. While Mom and Dad go diving, Vicki will teach the little ones to fish for snapper. Vicki's other specialties are the mangrove Back Country and the Lakes. She operates out of Key West Oceanside Marina on Stock Island.

Tate Berry is another wonderful personality who operates his huge barge type speed boat out of the Holiday Inn at the entrance to Key West. Tate is a highly experienced guide who knows all the Keys' secret spots.

There are two large sailing catamarans that currently take diving trips to the reefs around Key West: Tom Halford's *Chariot* and Bruce Robbins's *Chiquita.* Both are 36 foot (11 meter) Macgregor racing cats. They will get you to the reef as fast as many powerboats, and certainly in blessed silence. Halford sails from Land's End Village; Robbins leaves from the Pier House. Occasionally, I charter my own Warrior Racing Catamaran from the Pier House Motel.

At the present time there are only 2 major commercial operations that do dive trips to the Tortugas from Key West, Seasports at Land's End Village and Key West Pro Dive Shop on North Roosevelt Boulevard.

One other special trip that should be noted is run by Captain Billy Deans of Reef Raiders Dive Shop on Stock Island. Billy specializes in shipwrecks. His premier dive probably is the *Wilkes Barre,* a World War II ship in 145–250 feet (45–77 meters) of water in the Gulfstream. Billy's boat, the *Key West Diver,* is specially set up for deep diving with oxygen-decompression hookah gear.

The graceful flamingo tongue shell can often be found grazing on sea fans. The beautiful golden spots are not on the shell, but are actually the mantle of the mollusc inside. Photo: D. Kincaid. ▶

South Beach Patches and Key West Harbor 17

Typical depth range	:	15 feet (5 meters) reefs/30 feet (10 meters) harbor
Typical current conditions	:	Light on reefs but strong on harbor
Expertise required	:	Novice to intermediate on reefs, advanced in harbor
Access	:	Beach or boat

All along the southern shore of Key West (and in fact all the lower Keys), from 150–400 yards (45–120 meters) out, are scores of coral patches and reefs. Some are just individual heads, but others are extensive sets that cover many acres of ocean floor. The most prominent ones are off the foot of Duval Street, Simonton Street, immediately off of the Casa Marina Hotel, at the end of Bertha Street, off the main bathing beach, and off the airport. Probably the easiest ones to find are off the dock at the Casa Marina Hotel, where several markers can be seen. These mark a channel through the coral heads. Directly east and west of these marks are coral patches covering several acres of sea floor. Snappers and grunts are common as well as groupers and an occasional jewfish. In the grassy flats nearby are many eating-size conches and, of course, lobsters in the mud ledges and under coral heads. Almost any fish life that you might normally find on the

The South Beach Patches near Key West hold innumerable fascinating creatures such as this basket starfish. Photo: D. Kincaid.

A spotted moray wedges itself into a crevice beneath a variety of sponges on the South Beach Patches. Photo: D. Kincaid.

outside reef can be found here. There is seldom much current, and visibility is generally between 15 and 40 feet (5–12 meters), with the average about 30 feet (10 meters). All patch reefs along the Keys are in 12–15 feet (4–5 meters) of water and are found near shallow ledges or changes in depth. It has been speculated that they are probably the remains of Florida's ancient shore line. Although these areas are an easy swim from shore, you should always have a float or innertube with a diving flag prominently displayed, since on weekends and near the end of the day there can be a lot of boat traffic in these areas.

Advanced Divers. At the north end of Simonton Street is a small public boat ramp and dingy landing area which gives access to Key West Harbor. This harbor has been in use since the middle of the sixteenth century, and consequently the bottom is loaded with old bottles and artifacts as well as lobsters and grouper. Even though you have beach access here, a tank should be used in the 20–30 feet (6–10 meters) depths because of heavy commercial boat traffic. Visibility is about 15 feet (5 meters) with an extremely heavy current. This area is for *advanced divers only.* One favorite trick is to check where the cruise ships have been after they have left, since their prop wash can stir up the bottom, revealing the goodies. If you have a boat, a drift dive may turn up Spanish olive jars or bottles from the nineteenth century.

Sand Key 18

Typical depth range	:	Awash to 30–65 feet (10–20 meters)
Typical current conditions	:	Light to moderate
Expertise required	:	All levels
Access	:	Boat

Sand Key was originally called Cays Arena by the early Spanish explorers of the Keys. It is a simple sand island without vegetation, topped by a red iron lighthouse that was built in 1853 and is now on the historical register. According to the Spaniards, the island was "changeable according to the rigors of the weather." Indeed, with every winter storm or summer hurricane the shape of Sand Key changes. Upon close examination, the island turns out to be less sand than ground-up coral and small shells. Anyone wanting a collection of miniature shells can certainly find them here.

Since it can be seen for miles, the Sand Key lighthouse is easy to find and probably the most popular dive spot in the lower keys. It is not unusual to see several dozen divers and snorkelers here at a time. Yet there is never really a crowd because of the size and diversity of the reef and lagoon areas.

The lighthouse and beautiful shallow reef formations of Sand Key are clearly visible from the air. Photo: D. Kincaid.

A paradise for divers and snorkelers, Sand Key offers stands of elkhorn coral in very shallow water. Photo: D. Kincaid.

The reef itself is typical of major lower Keys reefs, consisting mostly of rock fingers and gullys of 5–20 feet (2–6 meters) deep, with sandy bottoms between cliff-like structures. Several sections of the reef have rather extensive areas of staghorn and elkhorn coral, and fire coral abounds. The northwest side of the reef has numerous coral heads and mixed rubble in close proximity to the lagoon and is more suitable to beginners. The south side of the reef gradually slopes away to a gentle ledge at about 65 feet (20 meters), dropping to 90 feet (28 meters) in some places. Also in the shallows are artifacts from the brick lighthouse, which blew away in 1846. Visibility at Sand Key can be highly variable depending on the wind, wave action, and monthly tide variations. At its worst it can be about 15 feet (5 meters) and at its best over a hundred feet (30 meters), but the average is 30–60 feet (10–20 meters). Spring and summer are the best times to visit Sand Key because the nearby Gulfstream blows in over the shallows, adding an extra sparkle to the saturated oranges and yellows that dominate the reef structure.

Outside Reefs 19

Typical depth range : 40–210 feet (12–63 meters)
Typical current conditions : Moderate
Expertise required : Intermediate and advanced
Access : Boat

For the more experienced diver, a dive on the outside reefs is a must. All along the Keys, just south of the main shallow reefs, are the Gulfstream reefs. These are the last diveable reefs in America before you get to the open sea. Also known as the Outside Bar, the Hump, 10 Fathom Bar, or Eyeglass Bottom, they are usually about a half-mile to a mile (1 to 1½ kilometers) into the Stream off the main reefs. Since they are all similar, I will only discuss one here; the rest are all on standard nautical charts. A few minutes' run south from Sand Key, the water starts to shallow up again, reaching about 40 feet (12 meters) at its shallowest, with a white sandy bottom. 80 percent of the time the visibility is 80–100 feet (25-30 meters) so you can just look over the side or use your fathometer; but when you get to 65 feet (19 meters), that's the drop-off and the last place you can anchor. The area above the drop-off is a lush prolific gallery of rarely seen deepwater corals and fish, reminiscent of deep walls in the Caymans and Bahamas. Over the drop-off the light is dimmer and there is not as much growth. The base of the reef varies from 110 to 210 feet (33–63 meters) where a gradual slope of rubble leads toward the ocean abyss.

The ledge of the dropoff along the Outside Reefs is typical of Caribbean formations, offering large heads of brain and star corals along with deep water gorgonians. Photo: D. Kincaid.

A nudibranch clambers over encrusted coral rock along the Outside Reefs. The animals are also called sea slugs, and come in various bright colors. Photo: D. Kincaid.

One of my favorite sightings was a 500-member school of giant tarpon that ranged from 7 to 12 feet (2–3 meters) in length (I have witnesses!). Since the full life cycle of the tarpon isn't known, this school of record breakers was a real treat.

Probably the best way to dive here is to start in the deep water and work your way up to the 40 foot (12 meter) level as partial decompression, then move back to Sand Key for another dive. There are several places along the wall where encrusted telephone cables spin off into the blue mist to Cuba, sponges are big enough to stand in, and black coral is shrouded with carousing Lilliputian life forms.

The current is occasionally strong but usually moderate, and water temperature even in winter is usually above 75°F (24°C) thanks to the Gulfstream. The deep reefs are different from anything anywhere else in the Keys. When you dive there and look to the deep sea, you are on the last bit of ground in America and have the entire rest of the U.S.A. at your back.

Rock Key and Eastern Dry Rocks 20

Typical depth range	:	5–35 feet (2–11 meters)
Typical current conditions	:	Light
Expertise required	:	All levels
Access	:	Boat

Rock Key and Eastern Dry Rocks are just east of Sand Key Light, about a mile (1½ kilometers) and a mile and a half (2½ kilometers) respectively. Both are typical of most reef formations in the area, with a rubble zone on top and long fingers of coral with sand and coral-filled canyons in between. Both are very popular dive spots with lots of shells, conches, lobsters, and fish, but their real claim to fame, besides their lush beauty, are the wrecks that are here.

Each reef is topped with rusted poles and both usually have breakers in all but the calmest weather. Depending on the wind, the best anchorages are usually on the western ends where there is generally less swell and a sandy bottom, since the battle between anchors and coral is usually lost by the coral.

Lobsters can often be taken while snorkeling. They like to hide in the crevices under coral stands on shallow reefs. Photo: D. Kincaid.

A diver with an underwater metal detector searches patiently for the remains of a Spanish wreck. Photo: D. Kincaid.

On top of Rock Key, in about 6 feet (2 meters) of water, are a number of large cement beams. In this area just beneath the sand lies a huge quantity of pebble ballast; since few people collected rocks in the 18th and 19th centuries, this area is a mecca for lapidary workers. Many of the rocks are quartz and there are several types of volcanic stone that can be polished to perfection and made into belt buckles, pendants, and so on. There are also a lot of tiles that say "Barcelona," brass spikes, and cannon balls.

On the western end of Eastern Dry Rocks in the second finger gully, just before the water is shallow enough for a person to stand up, is the remains of another 19th century wreck site. With ballast stones covering the bottom for over 100 yards (90 meters), anyone with patience or a metal detector can just fan away with a ping pong paddle and be rewarded with brass spikes, cannon balls, or an occasional bit of rigging. Since the average wreck yields over 250,000 artifacts, you can fan around for a long time, have a lot of fun, and never make a dent in the site. Most of the dive boats from Key West stop at these two reefs, so if in doubt just ask the captain for directions.

Western Dry Rocks (The K Marker) 21

Typical depth range : 5–120 feet (2–36 meters)
Typical current conditions : Light to moderate
Expertise required : Novice to advanced
Access : Boat

Just west of Sand Key, about 3 miles (5 kilometers), is a fine reef called Western Dry Rocks or the K Marker. The K Marker is not dived as often as Sand Key or other popular reefs closer to Key West, and consequently has not suffered the inevitable anchor damage that other reefs have. It is just a little further away than most weekend divers like to go, but is well worth the trip. Unless there has been severe weather recently, K Marker usually has clearer visibility than Sand Key, Rock Key, or Eastern Dry Rocks; reefs that straddle the major natural and shipping channels out of Key West.

Marine Life. The formation of Western Dry Rocks is typical of most of the lower key structure reefs: long fingers of cliff-like corals with sandy gullys in between. The K Marker has more cracks and crevices and many more caves than any other reef here. Species normally found in the Bahamas or on the deep outer reefs may be found. Primarily Candy basslets, Orange back bass, and an occasional pygmy angel fish or long snout butterfly. This reef is also the last of the reefs to the west with water shallow enough to have elkhorn and staghorn coral in any quantity before the Tortugas. The best anchorage is on the large sand bars at the western

The Western Dry Rocks offers a series of coral undercuts and ledges in water shallow enough for snorkelers. Photo: D. Kincaid.

Large marine life, such as this sea turtle can be found hiding in the caves at the Western Dry Rocks. Photo: D. Kincaid.

or northwestern end of the reef, or behind the reef rubble zone on the north side. The top of the reef can be waded at low tide, but is definitely not for non-swimmers. There are several large stands of coral on top of the reef and much debris from former reef markers and wrecks. These chunks of debris usually have lobsters. At certain times of the year, as many as 30 sharks can be seen patroling the top of the reef. This is evidently part of their mating ritual. If you see them, stay in the deeper water near your boat and *don't* spear fish or catch lobster; sharks will definitely come to the dinner bell and at these times they don't have good manners. Usually they are black tips and bull or lemon sharks, but an occasional oceanic hammerhead might be seen.

In several of the coral gullies of this reef you may find wreck material including lead sheathing, spikes, nails, and ballast. Many ships have struck this reef during the last couple of centuries. One crack in the reef even contains ballast stones embedded 8½ ft.(3 meters) under solid coral. At a growth rate of ¼ inch (6 millimeters) a year, that makes this wreck 400 years old. Most debris here is about middle to late 19th century.

Large jacks and permit can also be seen, usually in the spring and early summer. Water depths range from 15–25 feet (5–8 meters), but there is a gradual slope to 120 feet (36 meters) with sponges, coral, sea fans, and huge grouper. Water temperatures are just a little cooler here, but at the coldest only about 70°F (21°C). Visibility can be exceptional. I once anchored in what I thought was about 60 feet (18 meters) of water and my anchor took out 150 feet (45 meters) of line before it touched bottom.

Chet Alexander's Wreck I 22

Typical depth range	:	Awash to 30 feet (10 meters)
Typical current conditions	:	Moderate to strong
Expertise required	:	Novice to intermediate and advanced
Access	:	Boat

Local commercial salvor Chet Alexander had probably sunk more ships in and around the lower keys than any other individual—a destroyer escort, several barges, and a tug boat, to mention a few. His most popular dive site is a destroyer escort, locally known as Alexander's Wreck, that Chet bought from the Navy for $2,000 and moved to an isolated location.

The wreck is a popular dive spot, but is only occassionally visited by commercial dive boats. If you want to dive there commercially you have to ask. The wreck lies due west 5 miles (8 kilometers) off the #4 bouy near the Gulf entrance to the Northwest channel and about 3 miles (5 kilometers) north northwest of Little Mullet Key in about 25–30 feet (8–10 meters) of water. Alexander's Wreck is broken in half, with the stern section lying 150 yards (135 meters) or so north of the bow, which is awash on most tides. Despite the fact that part of the hull is clear of the water, the wreck is difficult to see. A number of captains have hit the jagged metal tear that juts above the water, ruining props and rudders.

A diver closes in for a macrophotograph on the crumpled steel hull of Chet's Wreck I. Photo: D. Kincaid.

A pair of spadefish hover on the sand near Chet's Wreck I. Photo: D. Kincaid.

Marine Life. This former naval vessel lies on its side and is home for thousands of fish. At different times of the year, different fish life can be found. Sheepshead, spadefish, pork fish, groupers, hog fish, snappers and angels are thick, with an occassional mackerel or jewfish making a visit. Spearfishers love this wreck. The hull itself is covered withLeafy oysters andJewel boxes, and occasionally Cowries orSpiny oysters can be found in the gun turrets.

Visibility is not quite as good on the Gulf side of the wrecks as on the Atlantic, but this area will usually average about 25–40 feet (8–12 meters). In the winter, water temperatures will be between 69 and 74 degrees F (20–23 degrees C), but even in the summer when the water temperature is in the 80s (about 28–32 degrees C) a wet suit and gloves should be worn, simply because of the jagged metal on the wreck and the stinging hydroids that cover the hulls.

At times the current can be very strong in this area, but generally not more than most divers can handle. Even novices can take great delight in snorkeling over the 2 parts of the site since the wreckage is generally visible from the surface. It is usually best to anchor away from the site and trail a safety line with a float on the end for returning to your vessel. The bottom in this area is covered with gorganians and sponges. Always send the first diver in the water to check the anchor's set because of the hard and relatively flat bottom.

The Aquanaut 23

Typical depth range	:	75 feet (28 meters)
Typical current conditions	:	Moderate to strong
Expertise required	:	Intermediate to advanced
Access	:	Boat

One of Chet Alexander's other dive sites is the wreck of the tugboat *Aquanaut.* Chet's former workboat sank at the dock; when he raised it, he determined that the boat was no longer useable as a salvage tug. It was taken to drydock and stripped of all useful items, then towed to sea and allowed to sink just outside the reef on a flat sandy bottom where nothing grew, a perfect artificial reef.

This 55 foot (17 meter) wooden salvage tug now sits upright in 75 feet (28 meters) of water on the edge of the Gulfstream, only a few hundred yards from the J marker "30" on Western Sambo reef. The tug is intact and in nearly perfect condition. Divers can enter the wheelhouse and crew's quarters forward, but it is not recommended to enter the aft compartment because of siltation and many hanging wires and debris. Since the boat is wood and not much subject to wave action at that depth, the hull and decks will eventually be eaten by teredo worms and collapse. It is reasonably delicate and might collapse at any time if disturbed, so be careful.

A school of glasseyes make their home on the Aquanaut, *also known as Chet's Wreck II. Photo: D. Kincaid.*

A small wooden tug, the Aquanaut *was towed to this spot and scuttled by local diver Chet Alexander. Photo: D. Kincaid.*

Marine Life. The wreck hosts a variety of fish life, including Mahogany snappers and Glasseyes as well as Spiny oysters. Generally the wreck is visible from the surface, and it is a fantastic sight sitting on the bottom, shrouded in fish life. When on the deck of the ship a macro close up lens is very useful. There are plenty of arrow crabs and deep water tropicals that are seldom seen by divers at other sites. It is a simple dive but a rewarding one for the experienced diver and always the possibility of big ocean fish materializing from the blue infinity of the Gulfstream.

Only one dive operator visits the *Aquanaut,* Captain Billy Deans, who works out of Reef Raiders Dive Shop on Stock Island.

Cottrell Reef* (Gulf Side Reef) 24

Typical depth range	:	3–15 feet (1–5 meters)
Typical current conditions	:	Light to moderate
Expertise required	:	Novice or intermediate
Access	:	Boat

Cottrell Reef is considered by most local divers to be an alternative dive spot when the weather blows up on the Atlantic side of the Keys. It is a shallow (3–15 feet or 1–5 meters deep) rocky ledge, protected from strong east southeast and southwest winds by miles of grassy banks called the Lakes. The ledges and banks of Cottrell start just west of the old house on stilts at the Gulf entrance to the North West channel. These ledges and solution holes are covered with gorganians and sponges, and run for a couple of miles before finally disappearing beneath the sands.

A photographer examines the cobbled surface of a coral head near Cottrell Reef. Photo: D. Kincaid.

A Lima clam and a convict goby enliven this patch of encrusting orange sponge on Cottrell Reef. Photo: D. Kincaid.

Marine Life. Intermittently along the ledge are large clusters of coral heads. The top part of the reef is flat and ideal for waders, but just a dozen feet (4 meters) away are pits, crevices, and coral caves carved by wave action. The bottoms of most solution pits are natural catch basins for shells. Weaving about these convolutions are a wide variety of juvenile reef fish, as well as many adults, mostly parrots and snappers.

According to Betty Bruce, historian at the Key West library, Cottrell Key was named after the Captain of the lightship that was anchored nearby in the early nineteenth century. The visibility at Cottrell is generally 15–40 feet (5–12 meters), and even on rough days there is seldom much current or surge. Consequently Cottrell is an excellent reef for beginning snorkelers. All of the commercial operations in Key West make this reef part of their itinerary when the weather is bad on the main reef.

The Lakes 25

Typical depth range	:	5–30 feet (2–10 meters)
Typical current conditions	:	Light, except in channels where it can be quite strong
Expertise required	:	All levels
Access	:	Boat

Directly west of Key West is a fascinating and varied diving area known as the Lakes. A series of grassy flats and banks completely encompass a shallow lagoon that starts at Mule Key and runs west 9 miles (14 kilometers) to Boca Grand Key. This shallow area is seldom more than 10 feet (3 meters) deep, and is usually less than 6 feet (2 meters). It is completely protected from wave action by the flats, and usually has reasonably clear water even in the worst winds. The area boasts several modern wrecks, coral and lobster filled channels, coral heads, sponge bars loaded with groupers and snappers, and islands with white sandy beaches. Large expanses of the Lakes are shallow and impassable, so if you go in your own boat, it is best to use a local guide or perhaps acquire topographic maps of the area, since regular nautical charts are of moderate assistance here.

The best known diving spots in the area are two target ships just north of Boca Grand Key. A large, deep channel on the west side of the island leads right up to the wrecks, one of which sticks out of the water and can be seen for miles. The other is submerged about 100 yards (90 meters) south in a finger channel. Both wrecks are covered with edible oysters and

The Lakes are a series of shallow lagoons protected by a string of islands and reefs. Photo: D. Kincaid.

Nudibranchs—which means "naked gills"—are so called because their gills are exposed. The graceful ruffles along this nudibranch's back are its breathing organs. Photo: D. Kincaid.

clouds of snappers, grunts, and an occasional school of snook (don't shoot, it's against the law), as well as an abundance of sea urchins.

The current on an outgoing tide can be much too strong for most people to swim against. Tide charts and good timing are a must, of course. If you wish to dive any of the channels in the Lakes at just any time, make it a drift dive, which covers more territory anyway.

Several islands in the Lakes have the letters *m u l e* in their names. This is left over from the ancient Spanish name for the islands, *Chici Mulei* (little brothers). These are some of the oldest place names still in use in the United States, dating back to the early sixteenth century. There are over 2 dozen major channels in the Lakes and some 40 sponge bars, with dozens of coral heads—far too many to explain here. With a good pair of polaroid glasses, a couple of charts, and a shallow draft boat, the astute diver should have no trouble exploring these fascinating mangrove islands. Don't forget to look beneath the mangrove roots; under these islands is a hanging garden of colorful life, unparalleled in its diversity, with numerous colorful sponges and macrolife. It is also a nursery area for almost all of Florida's commercial species.

Marquesas Keys 26

Typical depth range	:	5–30 feet
Typical current conditions	:	Light on coral heads; moderate to strong in channels
Expertise required	:	All levels
Access	:	Boat

The Marquesas Keys are the only known atoll in the Atlantic Ocean. Unlike their Pacific counterparts, which are volcanic in origin, these skull shaped islands are quite possibly the remains of a prehistoric meteor crater. The islands were named in 1623 for the Marquis de Cadierata, commander of the ill-fated 1622 Spanish treasure fleet, which included the famous wrecks of the *Atocha* and *Santa Margarita.*

The circle of islands is about 3½ miles (5½ kilometers) across and 22 miles (32 kilometers) west of Key West, or about an hour's run by outboard, and like most of the Lower Keys is a bird sanctuary and National Wilderness area. The Markeys, as they are called by locals, contain the only frigate bird rookery in the United States, and several of the islands have long, white, sandy beaches where pottery dating from 1622 can be found. There are also excellent anchorages in the various creeks between

A crinoid, or feather star fish, ducks into a small hole in the reef the Marquesas Key. Photo: D. Kincaid.

Delicate Christmas tree worms cover the surface of this coral head. Photo: D. Kincaid.

the islands. However, much of the lagoon is shallow and relatively impassable. There are a number of wrecks in the area, all of which are accurately placed on nautical charts. Huge clusters of coral heads can be found about 300 yards (270 meters) off the entire southern edge of the islands, in about 8–12 feet (3–4 meters) of water. Some of these are shallow enough to stand on and are also marked on the charts. The wrecks in this area are famous for jew fish, permit and cobia, but the coral heads are most noted for grouper, snapper and lobster. Although the Markeys are a good run from Key West, if you have the time and the gas they are worth it. They are only visited by serious divers and fishermen; the islands show little evidence of man and probably look very much as they did in 1622.

Although it's only about 5½ miles (8 kilometers) from the protected Lakes area, the crossing of Boca Grande Channel can be quite rough if the wind and tide are in opposition. Be sure to select your weather, spare parts, and radio with care and all the considerations of good seamanship.

Wrecks in Marquesas 27

Typical depth range	:	15–20 feet (5–6 meters)
Typical current conditions	:	Moderate to strong
Expertise required	:	All levels
Access	:	Boat

West of the marquesas lie several wrecks of interest to divers. The first is a little more than a half mile (1 kilometer) west of the westernmost cut of the Markeys, sometimes called Houseboat Cut because of the Treasure Salvors house boat that was anchored in the cut for a couple of years. This wreck was a target ship for the Navy many years ago and lies in a north-south position astride the first *major* sand bar off the cut. Parts of the wreck are shallow and just below the surface, so don't break your propellor here; it's a long way to the nearest repair shop. Large chunks of jagged metal lie in a ship-shaped pile, along with many old bomb casings. The sand bar shifts constantly; at times the wreck is completely covered, at other times well exposed.

About 3½ miles (5½ kilometers) further west lies another target vessel, a destroyer escort called the *Patricia* target ship. The wreck is easy to find because it is surrounded by a dozen large I-beams that stick out of the water about 15 feet (5 meters). This ship is still actively bombed by the Navy and Marines, so if a military jet makes a low pass over the ship, that's your signal to leave the area—firing will commence shortly and you must move off at least a mile (1½ kilometers) or two. Anchor and have a snack because the show can be spectacular.

The jewfish is gigantic member of the grouper family. Generally, they are a minimum of four feet (1 meter) long when fully grown, and jewfish up to six feet (2 meters) long and 600 pounds are not uncommon. Photo: D. Kincaid.

Some of the wrecks in the Marques are shallow enough for snorkelers to enjoy as well.
Photo: D. Kincaid.

Fish Among The Wreck. The *Patricia* is about 150 feet (45 meters) long and is now reduced to piles of jagged metal. Most of the rockets that are fired are practice bombs containing little more than shotgun shells, so they seldom do much damage and the fish life is unaffected. Although the current can be severe, it is usually easy to get behind a large chunk of the ship to calmer water. You should always wear gloves and some body protection when diving these wrecks because of the jagged metal and stinging hydroids. It is not recommended to attempt any penetration of the *Patricia* target; it is an extremely unstable wreck because of the bombing. Its configuration changes constantly, with large sections collapsing all the time.

If all this shooting business sounds like an outrageous way to go diving, take heart. The *Patricia* is seldom bombed more than a couple of days out of the month and is only usually worked over every three months or so. Bombing times and dates are listed in the newspapers and are on the radio. If in doubt call the local Navy base.

Cosgrove Shoal 28

Typical depth range : 20–210 feet (6–65 meters)
Typical current conditions : Moderate to strong
Expertise required : Intermediate to advanced
Access : Boat

About 6 miles (9 kilometers) south of the western side of the Marquesas, a 50 foot (15 meter) skeletal lighthouse marks the edge of the drop-off in this area and the northern edge of the Gulfstream.

Called Cosgrove Shoal after a nineteenth century captain of the lighthouse tender, this rocky bank runs for miles east and west and is top-notch diving for all activities, probably the closest any of us will get to truly virgin diving areas. Swimming around the lighthouse itself is a treat, because there you will see the legendary monsters of the deep—the Great Barracuda. These barracudas aren't the 4½ foot (14 meter) run-of-the-mill variety—some of them are in the 5 to 7 foot (1½–2 meter) range. Although there are usually only a half dozen of the big ones, the rest of them (often about 200) are in the standard 3–5 foot (1–2 meter) category. No matter how much you know about barracudas being harmless, diving among these will really get your heart beating.

The legs of the lighthouse on Cosgrove Shoal are covered with marine life. Here, wrasse and a red lipped blenny dart among seafans and encrusting sponges that coat the metal structure. Photo: D. Kincaid.

The claws of the Florida stone crab are a great delicacy. While restaurants fetch $12 or more for a plate of claws, divers can fetch their own at Cosgrove Shoals. Photo: D. Kincaid.

The reef that the lighthouse sits on is a prehistoric dead reef in structure, but it is riddled with coral caves and ledges that contain a wider variety of life than is normally found in the shallow elkhorn and staghorn forests of the Key West reefs. It has colorful crinoids, sponges, sea fans, coral heads, and every variety of deep reef life imaginable. The shallowest part of the reef is close to the lighthouse and about 20 feet (6 meters) deep; the slope goes both north to a rubble plain covered with sea fans at about 35 feet (11 meters) and out to the south in a gradual slope to 65 or 90 feet (20 or 30 meters), where it rises again to about 50 feet (11 meters) before finally dropping to 180–210 feet (56–63 meters). These deeper areas have occasional bushes of black coral. To my knowledge the only place in the continental United States where black coral can be found is on these deeper reefs. But since taking or possessing any live coral is against the law, admire its beauty and leave it alone.

The Gulfstream moves in regularly in this area and when it does, visibility can exceed 150 feet (45 meters). The average, though, is about 60–80 feet (18–25 meters). On an outgoing tide the current in this area can be severe, so dive upcurrent from your boat, trail a line from the boat with a float on the end, and always leave someone in the boat who knows how to run it in case your anchor breaks loose. Should that happen, the next stop south is Cuba.

Marquesas Rock 29

Typical depth range	:	20–120 feet (6–36 meters)
Typical current conditions	:	Moderate to strong
Expertise required	:	Intermediate to advanced
Access	:	Boat

Marquesas Rock is marked by a large can buoy and is only a mile and a half west of Cosgrove Shoal Lighthouse. There are numerous cracks and crevices in this large rocky plateau, and all around it are ledges that range from a couple of feet (1 meter) high to several sheer drops of 40 feet (12 meters) or more.

Marine Life. Huge schools of jacks and margates mass here with bait fish, squirrel fish, huge jewfish and turtles paddling about. Marquesas Rock is another place where truly wild wildlife will occasionally show

An enormous coral head stands out on the otherwise low relief of the reef at Marquesa Rock. Photo: D. Kincaid.

A southern sting ray, caught in a sand channel at Marquesa Rock, eyes the photographer suspiciously. Photo: D. Kincaid.

up—manta rays, sailfish, marlin, sperm whales and, believe it or not, "Jaws," the great white shark—or maybe just a 15 foot (5 meter) tiger shark.

Only a half mile (1 kilometer) north of Marquesas Rock is a wrecked vessel from about 1846. There are cannon balls, brass spikes, and rigging lying about, in about 35 feet (11 meters) of water on a shifting sand bottom at the end of a rocky ledge. The easiest way to find the wreck is to drag along the ledge behind your boat until you see three large water tanks and the rudder quadrant sticking up. There is also more recent wreckage mixed in with the older one, so it may have been salvaged in the early 1900s.

Again, you are about 30 miles (48 kilometers) from the nearest assistance, so be careful. Have lots of spare parts and a radio that works. The currents here are strong, but they are tidal and thus predictable. The animals are big, as is the sea, and your boat no matter how well found is small. It's a lot of water to drink.

6

Safety

This section discusses common hazards, and emergency procedures in case of a diving accident. For diagnosis or treatment of serious medical problems; refer to your first aid manual or the *Diving Accident Management Manual,* by Dick Rutkowski, published by the Florida Underwater Council. For information on obtaining a manual contact: Dick Rutkowski, Florida Underwater Council, 75 Virginia Beach Drive, Miami, FL 33149.

Diving Accidents. In case of a diving accident such as a lung overpressure injury (e.g., air embolism, pneumothorax, mediastinal emphysema) or decompression sickness ("bends"), prompt recompression treatment in a chamber may be essential to prevent permanent injury or death.

The Miami Recompression Chamber can be alerted by calling any of the following: Dade County Fire Rescue, 305 596-8576; U.S. Coast Guard Miami 305 350-5611; Radio HF 2182; Radio VHF Channel 16; or NOAA (VHF) Channel 9.

Emergency contact information can change unpredictably. Readers are advised to check on emergency contact information before going on their dive trip.

DAN. The Divers Alert Network (DAN), a membership association of individuals and organizations sharing a common interest in diving safety operates a **24 hour national hotline, (919) 684-8111** (collect calls are accepted in an emergency). DAN does not directly provide medical care, however they do provide advice on early treatment, evacuation and hyperbaric treatment of diving related injuries. Additionally, DAN provides diving safety information to members to help prevent accidents. Membership is $10 a year, offering: the DAN *Underwater Diving Accident Manual,* describing symptoms and first aid for the major diving related injuries, emergency room physician guidelines for drugs and i.v. fluids; a membership card listing diving related symptoms on one side and DAN's emergency and non emergency phone numbers on the other; 1 tank decal and 3 small equipment decals with DAN's logo and emergency number; and a newsletter, "Alert Diver" describes diving medicine and safety information in layman's language with articles for professionals, case histories, and medical questions related to diving. Special memberships for dive stores, dive clubs, and corporations are also available. The DAN manual can be purchased for $4 from the Administrative Coordinator, National Diving Alert Network, Duke University Medical Center, Box 3823, Durham, NC 27710.

DAN divides the U.S. into 7 regions, each coordinated by a specialist in diving medicine who has access to the skilled hyperbaric chambers in his region. Non emergency or information calls are connected to the DAN office and information number, (919) 684-2948. This number can be dialed direct, between 9 a.m. and 5 p.m. Monday-Friday Eastern Standard time. Divers should *not* call DAN for general information on chamber locations. Chamber status changes frequently making this kind of information dangerous if obsolete at the time of an emergency. Instead, divers should contact DAN as soon as a diving emergency is suspected. All divers should have comprehensive medical insurance and check to make sure that hyperbaric treatment and air ambulance services are covered internationally.

Diving is a safe sport and there are very few accidents compared to the number of divers and number of dives made each year. But when the infrequent injury does occur, DAN is ready to help. DAN, originally 100% federally funded, is now largely supported by the diving public. Membership in DAN or purchase of DAN manuals or decals provides divers with useful safety information and provides DAN with necessary operating funds. Donations to DAN are tax deductible as DAN is a legal non-profit public service organization.

Appendix

Dive Shops

*Indicates KADO Shop

UPPER KEYS:

Milemarker 120

The Dive Shop at Ocean Reef*
Box 15 Ocean Reef Club
Key Largo, FL 33037
(305) 451-367-3051
Tiny Wirs

Milemarker 106.5

Atlantis Dive Center, Inc.*
51 Garden Cove Drive
Key Largo, FL 33037
(305) 451-3020
Spencer & Amy Slate

Southbound Diving
45 Garden Cove Drive
Key Largo, FL 33037
(305) 451-1407
John Wallace

Milemarker 106

American Diving Headquarters
Rt. #1 Box 274-B
Key Largo, FL 33037
(305) 451-0037
Harry Keitz

Milemarker 104

Capt. Steve Klem Underwater Service
P.O. Box 1122
Key Largo, Fl 33037
(305) 451-1831
Steve Klem

Quiescence Diving Service, Inc.*
P.O. Box N-13
Key Largo, FL 33037
(305) 451-2440
Rob Bleser
Paul Caputo

Milemarker 103

Pennekamp State Park
P.O. Box 13-M
Key Largo, FL 33037
(305) 451-1621
Randy Pegram

Stephen Frink Photographic*
P.O. Box 19-A
Key Largo, FL 33037
(305) 451-3737
Stephen Frink

Milemarker 100

Sea Dwellers Dive Center*
P.O. Box 1796
Key Largo, FL 33037
(305) 451-3640
T.J. and Jan Holub

Mas Oro Dive Charters*
P.O. Box 69
Key Largo, FL 33037
(305) 451-1697
John and Judy Halas

Key Largo Marine Tours*
(at Holiday Inn)
P.O. Box 963
Key Largo, FL 33037
(305) 451-2220
Jerry Theiss
Jim Hendricks

Ocean Divers, Inc.*
P.O. Box 1113
Key Largo, FL 33037
(305) 451-1113
Joe Clark
Doc Schweinler

Milemarker 99.5

Capt. Corky's Divers' World*
P.O. Box 1663
Key Largo, FL 33037
(305) 451-3200
Corky and Terry Toth

Milemarker 90.5

Florida Keys Dive & Ski
90500 Overseas Highway
Tavernier, FL 33037
(305) 852-4599
Bud Gammon
Tom and Pam Timmerman

Milemarker 85.9

Lady Cyana Divers*
P.O. Box 1157
Islamorada, FL 33036
(305) 664-8717
Joan Follmer
Ken Wright

Milemarker 85

Reef Shop Dive Center
Rt. 2 Box 7
Islamorada, FL 33036
(305) 664-4385
Tom Tirb

Milemarker 84.5

Holiday Isle Dive Shop
P.O. Box 482
Islamorada, Fl 33036
(305) 664-1445
Ed Armstrong

Milemarker 79.5

Buddy's Dive Shop*
P.O. Box 409
Islamorada, FL 33036
(305) 664-4707
Buddy Brown

Milemarker 70

Scuba Doo Divers
Fiesta Key KOA
Long Key, FL 33001
(305) 664-8205
Chuck Aguilar

Milemarker 68.5

Atlantis Sea Ventures
105 Overseas Highway
Layton, FL 33001
(305) 664-4092
Don Gist
Dave Horne

MIDDLE KEYS

Milemarker 54

The Diving Site*
Coral Lagoon Resort
12399 Overseas Highway
Marathon, FL 33050
(305) 289-1021
Bob Tillman

Marathon Divers
12650 Overseas Highway
Marathon, FL 33037
(305) 289-1141
Jack and Rita Ferguson

Milemarker 53

Divers Unlimited
Marathon Inn Marina
Marathon, FL 33050
(305) 743-7262

Halls Diving Center*
Key Colony Beach Shopping
Marathon, FL 33050
(305) 743-0747
Bob Brayman

Milemarker 52-49

Diver's Headquarters
11511 Overseas Highway
Marathon, FL 33050
(305) 743-4501
Art McDermott

Ocean Equipment
2219 Overseas Highway
Marathon, FL 33050
(305) 743-4644
Wayne Quarberg

Hurricane Aqua-Center Inc.
4650 Overseas Highway
Marathon, FL 33050
(305) 743-2400
Ed Davidson

Milemarker 48

Hall's Diving Center and
Institute*
1688 Overseas Highway
Marathon, FL 33050
(305) 743-5929
Bob Brayman

Milemarker 31

Underseas Inc.
U. S. 1 Box 319
Big Pine Key, FL.
(305) 872-9555 or 294-0605
George and Maryanne Rockett

Milemarker 29.5

Keys Sea Center, Inc.
P.O. Box 515 MM29.5
Big Pine Key, FL 33043
(305) 872-2319
Capt. Bart Masker

Phil's Dive Shop
Cudjoe Key, FL 33402
(305) 745-3825
Phil Declue

Summerland Dive Shop
P.O. Box 321
Summerland, FL 33042
(305) 745-1890
Herb Karle

KEY WEST

Reef Raiders Dive Shop*
U.S. 1, Stock Island
Key West, FL 33040
(305) 294-0660
Capt. Billy Deans

Key West Pro Dive Shop*
1605 N. Roosevelt Blvd.
Key West, FL 33040
(305) 296-3823
Bob Holsten

Coral Princess Snorkeling
1990 Roosevelt Blvd.
Key West, FL 33040
(305) 296-3287

Island Divers
614 Green Street
Key West, FL 33040
(305) 296-4984
Jim Solanick

Key West Reef Trips
133 Duval St.
Key West, FL 33040
(305) 294-8383

Reef Queen Snorkeling
6 Duval St.
Key West, FL 33040
(305) 296-8865

Reef Raiders Dive Shop
Duval & Front St.
Key West, Fl 33040
(305) 294-3635
Capt. Franco

Index